THE IMPACT OF THE HUMAN STRESS RESPONSE

THE IMPACT OF THE HUMAN STRESS RESPONSE

The Biologic Origins of Human Stress

Mary Wingo, Ph.D.

COPYRIGHT©

First paperback edition, 2016

Roxwell Waterhouse
P.O. Box 1505
Austin, Texas 78767

Cover design by Rakib S
Cover art and book illustrations by Mary Wingo
Book design by Jun Ares
Editing by Heaven Stubblefield, and Bruce Wilkinson

ISBN 978-0-9974813-1-0 (Paperback)

CONTENTS

INTRODUCTION

WHAT IS THIS BOOK ALL ABOUT?

The purpose of this book is to explain one of the most complicated fields in science – how the human body responds to stress, and how that response leads to disease. The study of human adaptation ranges from the complex molecular level, to the cellular, organ, and to the whole organism, as well as to deep social, political, and economic issues.

The information has been outlined in a simple and readily understandable way so that an ordinary grandmother sitting in a swing on the back porch drinking sun tea can understand the concepts and add her wisdom to the discussion.

I believe all science should be communicated clearly. I also believe that the scientific concepts that the public has had a part in paying for over many decades should be explained in a way that can be understood and synthesized into informed decision-making and effective life choices.

Concepts in science should be framed in a manner that is easily understood by the general public. And if we simplify things in ways that policy makers, health care professionals, top level executives, and other influencers can understand and act upon, it will contribute to and solve many problems that plague advanced societies.

What I am going to set out is a radical insight into a much misunderstood issue. This opens the door to a whole new understanding of human disease and wellness, explaining how stress and the stress response affect us both for good and for bad. Specifically, I want to emphasize the importance for the general public with understanding basic concepts of personal biology.

How to effectively manage stress

Despite the crippling consequences of excessive exposure to stress, we do have many tools available to lessen these costs. These steps do not require money or expensive medical care, and can be used by anybody undergoing stress. Following these simple steps can result in a much higher quality of life and less resources wasted from dealing with needless stress. Millions of lives and trillions of dollars can be saved every year if we can address human stress on a global basis.

Step one

Go to chapter nine and find the five common sources of stress in our modernized society. Write down how many of these impact your life the most. For many living in the modern world, all five categories could be of impact.

Step two

Of those sources you have listed, create a separate list of specific stressors that are unique to you and your situation. For instance, if working memory capacity (chapter nine) seems to come up short, list all of the things you think about in a typical day, the multitasking you do, and the devices you use that exhaust your thinking capacity. Take some time with making your personal list -- it is likely to be quite long! You are simply taking inventory of all your stressors.

Step three

Commit to memory these rules about human stress:

- ✓ ***Stress is not only due to the threat or fear. Stress results because we attempt to adapt to our environment (chapter one). Disease occurs when this mechanism is overused or malfunctioning.***
- ✓ ***It is critically important not to add extra stressors to the already stressed body part during the Resistance Stage (chapter five).***
- ✓ ***The most intense stressors we are experience are between each other as social beings (chapter eight). Learning the skill of empathy is crucial.***
- ✓ ***Along with repair, maintenance, and reproduction, we also spend much of our energy adapting to the environment. We must take full inventory of our lives for the price we pay for always "adapting" (chapter two).***
- ✓ ***When the stress response mechanism becomes fatigued, the affected body part becomes exhausted. These affected parts can include the mind, as well as all other organs.***
- ✓ ***As we try to adapt to our environment, we must take note how much "novelty" we endure. If it is too much, we must***

limit our stimulation or our stress mechanisms will fatigue, limiting further attempts to adapt.

Step four

After you have made the long list of personal stressors and you have committed the above rules to memory, read chapter five. Do you tend to prefer routine or excitement in your day-to-day life? If you have a habit of craving too much excitement, you could open yourself up to stress related health and mental problems. For example, too much stimulation can overtax your working memory, leading to loss of judgment. However, too little stimulation can dull your creativity as well. It is important to have balance between the two.

If you have trouble identifying these traits, have someone who knows you well who will help you.

Step five

If you find yourself with a long list from step three, identify the top stressors on that list that you think take most of your energy. Eliminate as many has you humanely can. There are many you can reasonably remove.

You then have a list of stressors you cannot remove. Take a look at the other energy wasters on you list. Be very strict and assess and commit yourself to removing as many of the smaller stressors that you can. Small stressors are like dripping faucets, and the small drains add up to huge effects on your body and mind.

Step six

Now that you have addressed stresses you can directly control (this will take a while), take an assessment of the stress in your community. Do your neighbors, friends, colleagues, or other family members suffer from the effects of excess stress? Of the five stressor types discussed in chapter nine, which affect your community the most?

Step seven

If your community experiences a lot of stress, cultivating social capital (chapter nine) is an absolute must for reversing any damage caused. If you can form community groups, use this book for discussion and repeat steps one through five. This can also take patience and time, but the payoffs are huge! Developing social capital is the most important step a community can take to alleviate the health, social, and political problems present because of stress.

Step eight

If you want to create further positive change, discuss this book with community leaders, policy makers, and employers.

Effectively controlling public and personal stress results in longer life spans, better health, higher functioning, and more sustainable forms of human economics. With the simple explanations and solutions presented in this book, I hope to make a positive humanitarian impact during these times of worldwide uncertainty and turmoil.

This book is not exhaustive about the topic of human stress and adaptation – it is merely an overview. Many future volumes will be written about other important stress-related issues.

I hope you learn a lot from this book. As always, I am open to feedback and suggestions for future topics.

Mary Wingo, Ph.D.
Cuenca, Ecuador 2016
www.marywingo.com

CHAPTER ONE

UNDERSTANDING THE STRESS RESPONSE

Chapter Objectives

- ✓ ***Stress is not only due to the threat or fear. Stress results because we attempt to adapt to our environment.***
- ✓ ***The more we exchange energy and matter with the environment, the less we are able to defend ourselves from stressors within the environment.***
- ✓ ***Inflammation is a common reaction to stress. It helps insulate the effects of a stressor from the rest of the body.***
- ✓ ***If inflammation goes on for too long, cellular nutrient delivery and waste removal become difficult. This is one way disease forms.***
- ✓ ***The mechanism that controls our open and closed states with the environment can break down from misuse over time. This is another way disease forms.***
- ✓ ***Putting stress on an already stressed system ultimately results in no good.***

What Is Stress?

Specifically, what is human stress? This is a difficult question, and you will get different answers depending on who you ask. Personal stress can range from taking care of others to having a demanding boss to sitting in traffic. Dealing with a chronic disease, getting a divorce, serving in combat, or being in an abusive domestic environment - all of those situations can be considered "stressful." But just as importantly, the body also gets stressed from things that most people don't often think about, like maintaining a poor diet. Eating processed food, or not having enough of the right nutrients can cause symptoms of stress, even if they don't originate in the mind. As we will see in a minute, stress isn't just about how you feel; it is about how your body copes with a variety of challenges on many levels.

We each know from our experience of stress what it feels like, and the toll it takes on us. Nevertheless, it's difficult to create a concise definition of what is happening when these stressors affect us. (A "stressor" is anything that makes us stressed.) Even Hans Selye, the brilliant scientist-physician who coined the term "stress" could not express a precise definition of what stress is. His best attempt in 1936 was that stress is "the non-specific response of the body to any demand for change." Despite a lifetime of effort, he was unable to tighten the definition any further.

THE CONTEMPORARY DEFINITION OF STRESS IS:

"The rate of adjustment a cell, tissue, organ, organ system, or organism undergoes in order to adapt to a given environment."

Let's explore this definition a little further It is important to realize that a living organism's stress response is not just psychological. Physical parts of us can get mechanically or chemically stressed as well. A muscle cell in your foot can undergo stress when you are hiking, even though your mind is seemingly unstressed and admiring the nature.

The experience of stress is not just the result of traffic, a demanding boss, or a failing relationship. It is not just the experience of frustration or helplessness. Stress is the cost our bodies incur when adjusting to a particular environment.

An Unconventional Way to Look at Stress

Let's say you are a blues singer living in New Orleans, the Jazz Capital of the world. New Orleans is about sea level in altitude. Your cardiovascular and respiratory systems have adapted to belting out songs at the top of your voice at sea level. Then let's say you fly to Quito, Ecuador to give a concert. Quito is an Andean city in South America that is higher than 9,000 feet altitude.

Yesterday, you were singing at sea level, and today you are doing the same almost two miles higher. It is no stretch of the imagination to say that your heart and lung systems would be strained. I observed this myself on a trip to Quito, and saw the sickly-looking singers who had recently flown in having to use oxygen tanks to recover between performances.

Their respiratory distress was a good example of a couple of organ systems having to adapt to extreme differences resulting from changed altitude. It is also a good illustration that explains clearly that stress is not due to mental agony alone. Stress is neither good or bad-it is simply a demand from the environment to adapt. *Stress is due to the effort it takes to adapt to whatever environment you find yourself in.*

The Three Types of Stress

All living organisms can encounter three types of stress:

1. Psychological (for animals with advanced nervous systems)
2. Chemical
3. Physical

We will define these three types of stress in detail at a later time, but for now, you should know that psychological stress is the most potent and common of the stressor types that we face. It happens when the resources of the prefrontal cortex of our brain (especially the lower part behind the eyes) are insufficient to cope with a cognitive or emotional stressor.

With trying to cope, the rate of adjustment imposed by the emotional system of your brain exceeds the capacity of the prefrontal cortex. The "wear and tear" that results is called the *allosteric load*.[1] When this load is too high, we become "stressed."

The Chemistry behind the Stress Response

The chemical stress response first developed long ago when our microorganismal ancestors were faced with challenges while living within a fluid environment. The destabilizing effects of changing pH, energy, fluid levels, and other dynamic chemical properties all presented obstacles that they needed to overcome. If an environment is toxic, a microbe will detect the changes through a process known as *chemotaxis*, and attempt to move away.

More complex organisms, like humans, respond to chemical stressors differently because we are more likely to become stressed by an internal exposure (something eaten or breathed) or a problem with our internal chemistry. These are things we cannot escape from. Electrolytes, pH values, energy concentrations, blood gases, and other signaling molecules (hormones, neurotransmitters, antigens, and toxins) all have the potential to destabilize our body's equilibrium. This causes stress because our bodies have to adapt to the change.

All stressors - chemical or physical - cause primitive responses in humans. Too much sun will burn us, and too much cold will freeze us, but we can cope with a considerable amount of physical stress from exposure to the elements before it overwhelms us. We can also cope with a considerable amount of wear and tear, but if we overwork our joints and muscles, the next day we feel the painful effects.

Stressors Like to Hunt in Packs

In reality, we are often exposed to multiple types of stressors within a short time period. For instance, if you are on Spring Break in the Bahamas, you can experience chemical stress from imbibing too much beer and eating too much pizza. Your girlfriend may get disgusted by your debauchery and break up with you, leading to emotional stress. This could lead you to get even more drunk, resulting in more chemical stress. Then you pass out on the beach in the afternoon in the full sun and get the mother of all sunburns. Now your body has physical stress to deal with, on top of the emotional and chemical stress that you were already going through. Your vacation, which was supposed to be a time of rest and relaxation, has resulted in physical, chemical, and mental exhaustion that ultimately leads to cellular aging and loss of human functioning.

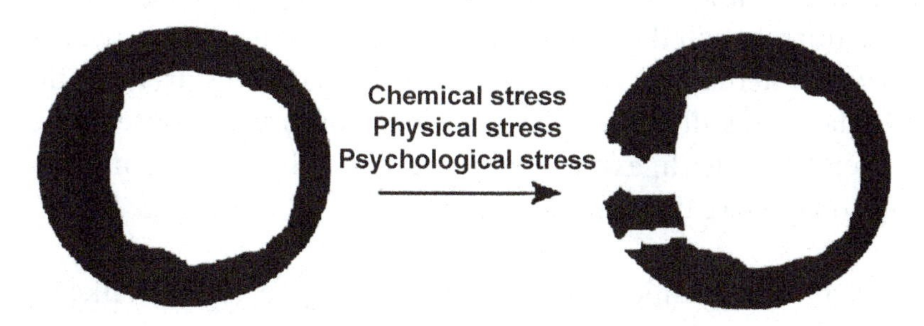

Multiple intense stressors (some of these can be fun) experienced within a short period of time leads to tissue breakdown and aging

There will be time to explore these stressors in-depth later. For now, it's enough to take on board that human stress is not just about fighting off threats to our existence. It is about coping with the demands of "adjusting" successfully to our environment.

Surprisingly, our environment is not just external. If we look at stress on a cellular level, we find that temperature, pH, blood sugar levels, etc. can sometimes be so badly out of "whack" that the adjustment that our cells have to make to survive creates a stressful environment. In this case, the "environment" is actually inside us!

> ✓ ***Stress is not only due to the threat or fear. Stress results because we attempt to adapt to our environment.***

The Boundary Hypothesis – Another Way of Looking at Stress

In looking at the huge topic of stress, we will first explore the human experience of stress, and then work our way downward. To help make things clear, let's shift our thinking a little.

We tend to think of ourselves as static, unchanging beings. We look in the mirror each day and expect to see an image that is the same as it was yesterday. We also expect that the hats and shirts that fit us yesterday will still fit us today. Our loved ones and spouses, whom we spend time with, expect us to look and behave pretty much the same today as we did yesterday or last week.

But what about reunions that take place many years apart, like high school class reunions or family get-togethers? Seeing someone after a long period apart makes us realize that we are not static – we are actually quite plastic. That is to say, we change. In fact, even though we may not notice it, we actually change morphology (our body shape) on a day to day basis.

This is not the entire story though. Think of an organism as a functional unit, like a machine. Or better yet, let's compare ourselves to an amoeba, whose shape and boundaries are flexible and ever-changing. Our system is part of the natural world, and therefore follows the laws of thermodynamics. This means that, being plastic as we are, we exchange energy and matter (really the same thing) across our borders with the environment every single waking moment.

So far so good?

Understanding that organisms are constantly exchanging energy and matter with the outside makes it much easier to understand stress. What actually happens when we are stressed is that our *mechanical stress responses* "adjust" our borders with the environment to become more "open" or "closed". This concept is called the ***Boundary Hypothesis***.

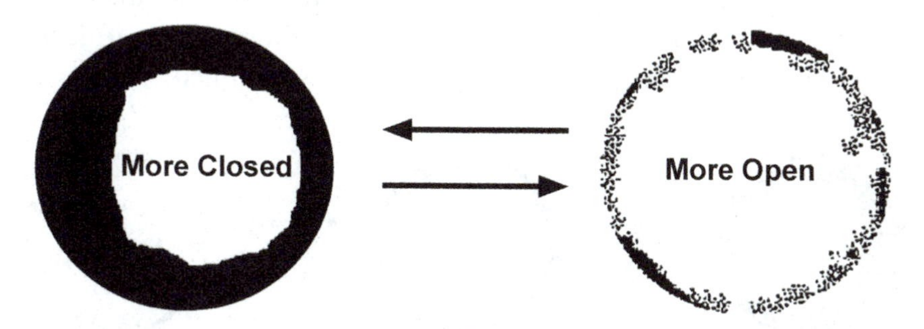

Our bodies are in a constant state of flux. We may think we remain the same from moment-to-moment, but in reality, we are in a state of constant change

The hormonal, immune, and nervous components of our mechanical stress response act deliberately to make us more plastic, and therefore more open to influences from the environment.

What do we mean by becoming more plastic and open to the environment? It means that we literally relax our borders on the cellular, tissue, organ, or organismal level, and *change ourselves to fit the environment*. In other words, we adapt to the environment not by keeping it out, but by letting it in. This allows us to survive as we let the environment become *part of us*.

This is exciting, because this tells us what stress is really about. **It is about being forced to adapt.**

Peter Csermely, perhaps one of the greatest complex systems thinkers in the world, has explored the problem of adaptation extensively. He has shown that complex systems of the self-sustaining sort (like us) undergo plastic and rigid phases to communicate effectively with the environment. This makes it easier for us to rearrange ourselves to adapt to new conditions.

As described by Csermely, an organism's plastic state is highly adaptable. The trade-off is that it loses "memory" of the overall system layout. Whereas in the rigid state, memory of structure and function is retained. But because of its inflexibility in responding to environmental signals, its ability to adapt to stressful conditions is limited.[2]

Our "borders" become more open or closed to the influences of the environment. The condition of both the environment and the organism affects the efficiency and speed at which these changes take place.

So we have two choices: either to be open and adaptable, or rigid and defensive. Our **stress response** controls how we alternate between these two physical states. This adaptive mechanism is not just present in human, but in all living creatures. It is a core feature of life. This concept, called the *Boundary Hypothesis*, is the primary take-away of this book.

Ideally, it switches us between the two at the ideal time and in the right amount to achieve a perfect blend of plasticity and rigidity. This gives us the adaptability we need to survive in the face of changes in our environment, plus the ability to retain memory of systemic processes (the "purpose"). We need this to keep our cellular processes organized. If there is too much plasticity, our biological processes risk becoming plunged into chaos.

Think about the Blues Singers...

Let's go back to our example of the blues singers. If you remember, they were going from sea level to over 9,000 feet. Bearing in mind what we have just discussed about being plastic and changing to adapt to our environment, it may now be easier to understand why the singers' tissues have to expend so much effort to adapt to the sudden increase in altitude.

Firstly, there are huge decreases in oxygen and barometric pressure. Then there is the added demand for adjustment that comes from the extra pressure they are putting on their heart and lungs by singing. Because human tissue takes time to structurally reorganize, the singers needed a stress response to keep from falling into systemic shock while performing. The stress response acts as a type of lubricant, preventing major organ systems from seizing up and malfunctioning during exposure to a stressor.

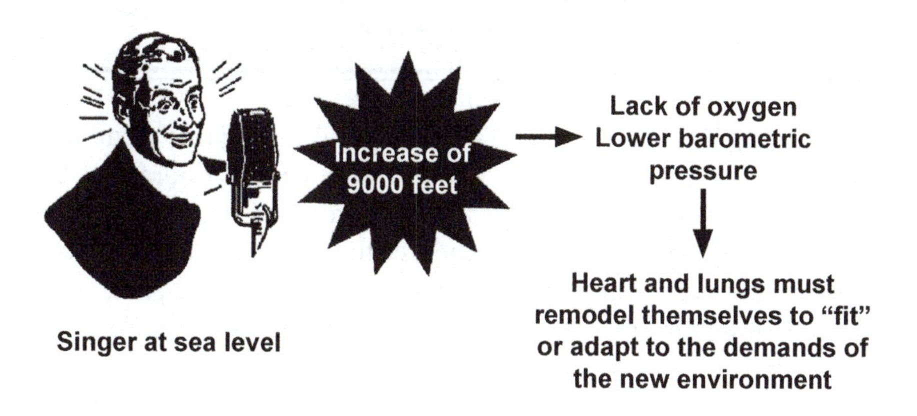

My observations showed me that these singers were struggling to adapt. They needed frequent rest breaks and an oxygen tank. It is likely that those with the most efficient and robust stress responses were the most plastic, adapted more quickly to their new environment, and gave the most effective vocal performances. But for others, the experience was too much. Singing old Dizzy Gillespie tunes is stressful enough at 9,000 feet. For some of these singers, who were older and not in good shape, the added pressure of being unfit made performing almost impossible.

The Cost of Openness and Biological Flexibility

We now can easily understand how biological flexibility gives us the upper hand with adaptation and survival in a changing environment. This ability, however, may not be all it's cracked up to be. Our bodily states are not just about being flexible and plastic. Remember the risk of biological chaos we mentioned a moment ago? Being physiologically flexible and more open to the environment comes at a high biological cost. That is because becoming more permeable causes an organism to lose structural organization and defensive ability.

> ✓ ***The more we exchange energy and matter with the environment, the less we are able to defend ourselves from stressors within the environment.***

For example, if we turned the Empire State Building into a rubbery sponge, we would not have much faith walking on its floors wearing stilettos. Or imagine the huge leaves on a palm tree. They whip effortlessly in the wind allowing the tree to survive undamaged even in hurricanes, but they need to be grounded by a compact, solid trunk. Without it, their flexibility would be wasted.

The same applies to us. When we are in a state of a classical stress response, we become more open to the environment, thus reducing our "grounding.

Palm tree fronds are constructed in such a way that makes adaptation to high winds possible

Cellular reproduction and repair come to a standstill. Many parts of our defense, such as the immune system response, slow down dramatically. Think about it: who needs defense if you are in a state of becoming more open to the outside environment? For even more curious reasons that we will go into later, in a robust stress response, inflammation and other aspects of immune response are minimized.

Illustrating the Stress Response

To illustrate what's going on here, it is as though the stress response relaxes the concept of what we think of as "self." The point to remember, therefore, is that the "self" as a system becomes less clearly defined when we are under stress.

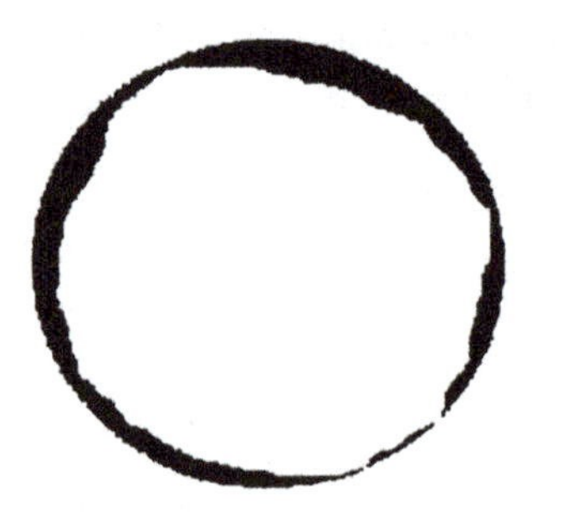

Self before the stress response **Self after the stress response**

The figures above represent an organism at rest and under stress. When the cells and nervous detect a stressor, the tissue first goes into a defense mode and then an adaptive mode. Notice that the boundary between self and non-self in not uniform or regular. Though this self-non-self boundary is composed of many anatomical and physiological aspects, an important core trait is basic fluid pressure being held outside the cells and organs in the space outside of the cell forms the actual shape of the body.

Inflammation is one of the first defensive response to a stressor and the increase in extra-cellular fluid pressure makes the affected area more rigid and firm to the touch. Effectively, the pressurized fluid forms a sort of straight-jacket that isolates the injured or stressed area from further environmental stressors.

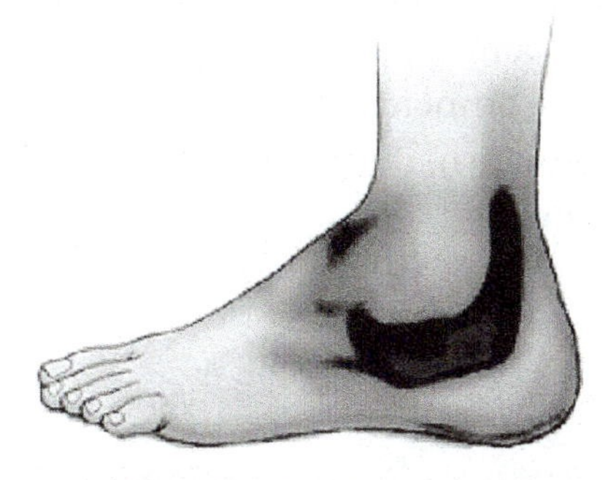

During inflammation (sprained ankle as an example), the organism exhibit many defensive behaviors:

1. Increased fluid pressure — hard, rigid tissue result in increased pain, increased heat, as well as an immune response

2. The swelling traps the invader or debris, not allowing it to move

3. The swelling causes pain, thus inhibiting movement

4. While clean up and remodelling is occuring at the cellular level, the ankle becomes more "closed" to the environment with the help of pain

5. The result of closing this body part off from activity is to lessen participation in the environment while the tissue reorganizes

Stress and Inflammation

A good example of the discussion above would be a sprained ankle. Immediately after an injury, the inflammatory response kicks in. Small blood vessels in the ankle become leaky, and release fluid in the form of plasma or lymph (as well other proteins and immune cells) into the area. Due to increased fluid pressure, the injury becomes hard and movement in the ankle soon becomes difficult, if not impossible.

That is the whole point of the response. As a result of the difficulty in moving the ankle and the pain signals released, further injury become far less likely. This entire physiological and psychological cascade of events results in the organism becoming more defensive and closed to the environment. Usually, because of the pain, the person withdraws from normal daily activities until the injury heals and the stressor subsides.

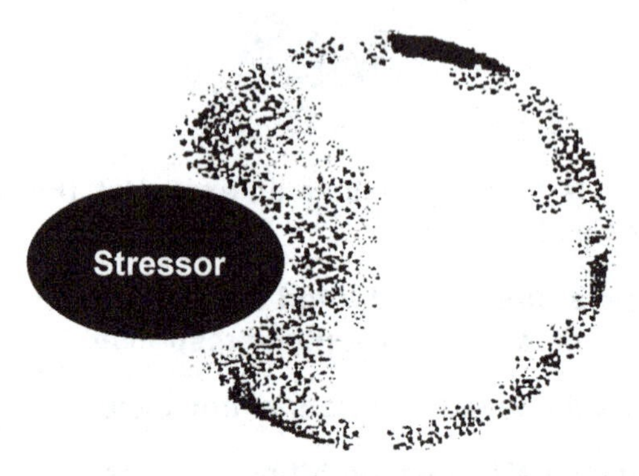

A successful stress response allows the organism to shape itself to the demands of the environment

Sometimes the stress – and therefore inflammation – is caused by microorganisms, such as a bacterial infection. In these cases, the fluid buildup and emergence of immune fighters and clean-up cells holds the infection in a type of biologic jail. In other words, the bacteria are head captive by fluid pressure and the immune help that come at a later time.

Under the process of inflammation, defenses go up, tissue becomes hard and fortress-like, and the materials allowed to pass in and out are limited.

Our body's border with the outside world is strengthened, and our sense of biologic "self" is reinforced. This boundary strengthening process keeps 'us' as 'us'. We need this response to let us repair, to absorb nutrients, and to start the process of reproduction. If we do not have an inner memory of our structure, we cease to be.

> ✓ ***Inflammation is a common reaction to stress. It helps insulate the effects of a stressor from the rest of the body.***

Stress and Tissue Response

Simply put, we as whole organisms have two physical states: one that molds us like clay to help us adapt to change, and another that builds a fortress like steel so we can grow and repair our sense of self. This arrangement also applies to individual organs and tissues as well.

We alternate between these two life states not only many times within our life, but also within each day. Our individual body parts alternate between these two states as well. These are ancient organismal responses that evolved side by side. If you were created without a stress response, it is unlikely you would survive the massive adjustment of simply being born. If a complex creature lacks a stress response when one is needed, it simply goes into systemic shock (the mother of all inflammatory responses) and dies from cardiac failure.

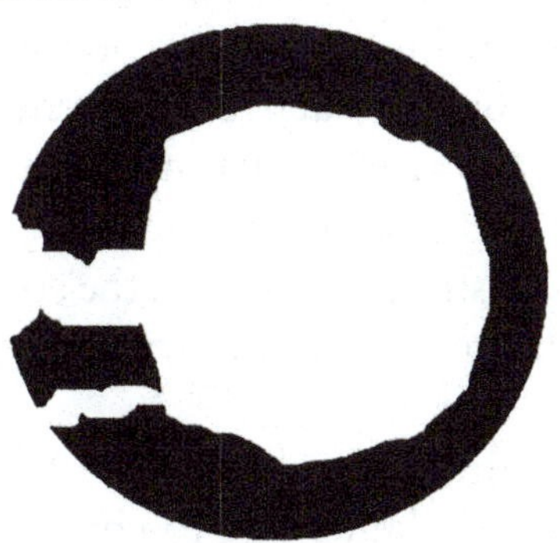

Tissue destruction and possible death occur when the stress response fails. This is due to tissue exhaustion.

When the defenses become too great and the borders to the outside become too closed, the organism fails to adapt to its environment.

If this maladaptive response does not do what it is supposed to do, the tissues, organs, or even the entire organism become exhausted and real functioning is compromised. The affected region becomes nonresponsive. If the exhaustive state continues unabated, the affected part (and even perhaps the organism) dies.

How does this happen? When your fluid transfers from inside of the plumbing of the blood vessels to the extra-cellular space, both nutrient and waste transfer drastically slow down.

At this point, one's cells dies from both starvation and exposure to its own toxic excrement.

> ✓ ***If inflammation goes on for too long, cellular nutrient delivery and waste removal become difficult. This is one way disease forms.***

Why Our Stress Response Sometimes Fails (and the Disastrous Consequences)

Like any process in our body, our two bodily states can become disturbed by disease of their delivery systems. For example, the adrenal gland is a major organ regulating the stress response. If this gland becomes inflamed and fails to secrete sufficient cortisol and adrenaline (both major stress hormones), the stress response will not be effective. This puts the organism's survival at risk.

People with the autoimmune disorder Addison's disease have malfunctioning adrenal glands due to chronic inflammation. In order to survive, adapt, and have a somewhat normal life, these patients must take the pharmaceutical drug equivalents of cortisol and adrenaline (cortisone and epinephrine) for the rest of their lives.

Disease of the stress hormone delivery systems is one possible problem, but a more common one in the case of the stress response is **maladaptive conditioning**. Maladaptive conditioning is when we make such great demands on these two opposing open-closed event states that the healthy regulation of stress and defensive responses break down. Stretching these systems beyond their natural capacity can culminate in almost any disease known to man. That is because while a healthy stress response protects us against disease, when the stress response breaks down, diseases can take hold.

Body begins life with a large ability to adapt → **Adaptive mechanism get overused (often over many years)** → **Body loses its ability to adapt and disease ensues**

> ✓ ***The mechanism that controls our open and closed states with the environment can break down from misuse over time. This is another way disease forms.***

Let's restate this, because it is so important. Disease in general is the natural result of overextending either or both of our two physical states. If that over-extension goes far enough, proper cellular (and thus organ system) functioning ceases. What happens before actual tissue death is that the healthy stress response fails to inhibit or "put the brakes" on runaway inflammation. This eventually leads to shock, or the launch of the dying process.

Keep in mind that this failure of adaptation happens on the cellular, tissue, organ, and whole body level. You can have individual organ failure without killing the organism, but it sure makes adapting to stressful demands a lot harder. It would be catastrophic if the blues singers we described earlier had severe heart disease to contend with while trying to belt out a Stevie Ray Vaughn tune at the top of their lungs at a very high altitude.

But again, if we think about this, it is pure common sense. **Putting stress on an already stressed system ultimately results in no good.**

Key Concepts:

- Identify and define the following concepts:

  Stress
  Three types of stressors
  The *Boundary Hypothesis*
  Inflammation
  Maladaptive conditioning
  Adrenal glands

- Identify common sources of stress that are present with you, your families and communities, as well as within the various levels of our society.
- Why does experiencing more than one stressor at the same time cause such big problems?
- What is the purpose of the stress response?
- What does being more "open" or more "closed" to the environment mean?
- What cellular and bodily functions shut down during the stress response?
- Why does inflammation occur?
- What happens to an organism if it cannot initiate a stress response to a stressor?
- How does a disease state begin?

CHAPTER TWO

WHO AM I? SELF VS. NON-SELF

Chapter Objectives

- ✓ ***The best way to understand complex organisms (like people) is to study the relationships between the parts, and not the just the parts alone.***
- ✓ ***Along with repair, maintenance, and reproduction, we also spend much of our energy adapting to the environment.***
- ✓ ***There are 10 times as many friendly organisms as human cells living on and within us. These microorganisms help the body perform complex functioning. They are part of "us".***
- ✓ ***Point to consider---When does matter or energy become part of "you" after entering your body?***

Defining "Self"

Before looking deeper into the topic of stress responses, we need to take a slight diversion to ask a very important question: **What do we really mean by 'self'?**

Who are you? Or better yet, *what* are you? Though this question may sound absurd, it needs considering because it goes to the heart of this book. To understand the stress response properly, we need to see ourselves as a **complex adaptive system**, as Peter Csermely suggests. We will explore this further in just a moment.

Now let's take a cautious step forward. We can certainly reduce ourselves to that of atomic interactions. This is a form of *reductionism*. If, from a reductionist view, we have no self or *agency*, then were does a person's free will and sense of self come in? Where does individual choice take place? Sure, most of our lives are out of our own control, determined by forces of nature and of others exercising their free will. But we also have *some* choice in our lives, and this reality is poorly addressed in biology. We can choose to have tea or coffee in the morning, or choose to have nothing at all. Many of us can choose our friends and our mates. We can choose to do a fitness program, or learn new skills. You are choosing right now to read this book.

Kurt Goldstein, like Selye, was another father of modern medicine. He worked in the field of neurology. He discovered much about the human nature by treating and studying thousands of brain injured soldiers in World War I. The remarkable insight he gave us is that we must look at the behavior of the organism in a holistic, and not in a reductionist way. What he discovered can be summed up like this: ***When we start isolating specific parts of a complex system from that system, the parts behavior changes, because we have isolated that part's function from the bigger whole.***

What this means is that when feedback loops with the greater whole are broken, the isolated part that we are looking at changes. As a result, we get distorted information. Isolating any of these input methods from the whole is bound to give skewed, non-representative observations as described by Goldstein and shown in the figure on the right.

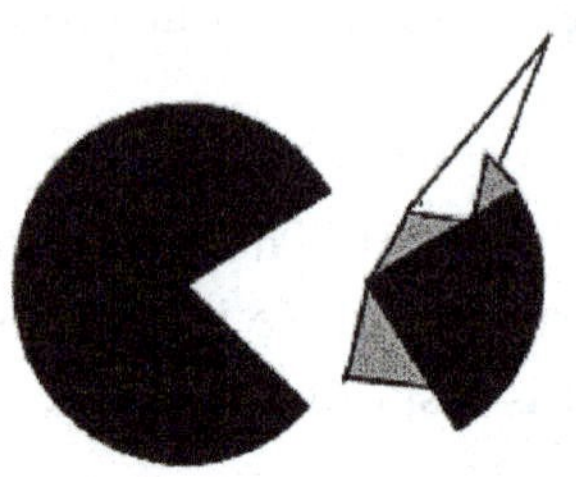

When isolated from the whole for obsevation, the nature of the part changes dramatically

If we heed the words of Goldstein, we realize the serious error in attempting to isolate and study a organism's part from the whole. This is because the reductionist's view only concentrates on the part of interest (like the heart or liver), and fails to take into account about *how* these parts work with each other and with the whole organism. We lose a tremendous amount of information employing this method of study. We must use a different approach for studying complex organisms.

> ✓ ***The best way to understand complex organisms (like people) is to study the relationships between the parts, and not the just the parts alone.***

To solve this conceptual problem, it might be fruitful to turn to the engineering brainiacs for assistance. Living organisms fall under a category of systems called the *Complex Adaptive Systems* (CAS). A CAS is not only complex, but it can shift itself under a changing environment. The qualities of CAS when applying it to human process come up as much more satisfying than the reductionist, trees-for-the-forest approach.

In order to understand Complex Adaptive Systems a bit better, let's look at some of their attributes. The qualities of a complex system include the following[3]:

- **Distributed Control:** No executive boss, control is dispersed. The whole is greater than the sum of its parts.
- **Connectivity:** Inter-connectivity between units within the system and with the systems outside environment/ecology.
- **Co-evolution**: As discussed previously, a perfect example is the mutual co-evolution with microbes in our gut, skin, and various orifices.
- **Sensitive Dependence on the Starting Conditions:** This is the reason why early childhood nutrition and experiences result in great significance for adult outcomes.

- **Emergent Order:** Because our system is so complex, patterns begin to emerge. Such examples are the patterns we see emerging from the general functions of digestion, nervous, or any other organ system.
- **Far from Equilibrium**: Chemical equilibrium is different harmonious, balanced, and healthy functions of life, working in a state called *homeostasis.* Chemical equilibrium occurs when substances, such as salts or sugars (or other components), are equally distributed within the space of the system, e.g. mixed thoroughly. Part of being alive is having unequal gradients, or concentrations of nutrients and other factors within our system.
- **State of Paradox:** Perhaps the most fascinating aspect of living systems. It is truly remarkable how we keep order going within our system. The default state for most of us is to keep up orderly biological functions. At the same time, however, when, adapting and adjusting to stressful conditions, we become more plastic We begin sliding into a *controlled chaos* situation until the need for adjustment subsides.

As you can see, looking at ourselves from a whole person perspective, our adaptive biological mechanisms make a lot more sense. We can conceptualize the human body as this undulating, ever-changing plasma-like structure. Our borders, however, as we may define them, are flexible, providing more permeability with our outside ecosystem when we need to adapt, and snapping back into a more protective, more rigid structure when we are without the need to adapt.

This new way of looking at ourselves clarifies the very disorganized description that our current biological perspective gives us. That is, the organism not only repairs and reproduces itself, it also spends a huge amount of resources adapting (or adjusting) to the environment.

> ✓ ***Along with repair, maintenance, and reproduction, we also spend much of our energy adapting to the environment.***

Microorganisms and "Us"

First, let's do some navel gazing. What actually constitutes "you" vs. "not you?" On the surface, it should be easy. You look at your hands and you see a *line of separation* that separates you from the air around your fingers.

However, the reality is not so clear if we realize that we are not simply stand alone units. For instance, the microorganisms that live on us and within us outnumber our own native cells ten-to-one, and make up about half a gallon (about two liters) of volume. It is estimated that about 37.2 trillion native cells makes up each core human unit, which means we have something of the amount of over 370 trillion non "us" cells living on and within us.[4-5]

To put this into perspective, this is more that 1000 times the number of stars in the Milky Way!

> ✓ ***There are 10 times as many friendly organisms as human cells living on and within us. These microorganisms help the body perform complex functioning. They are part of "us".***

These critters inhabit just about every surface and orifice we have, and they do so **symbiotically**. This means that our relationship with them is mutually beneficial. What do they contribute in return for our hospitality? They help us adapt.

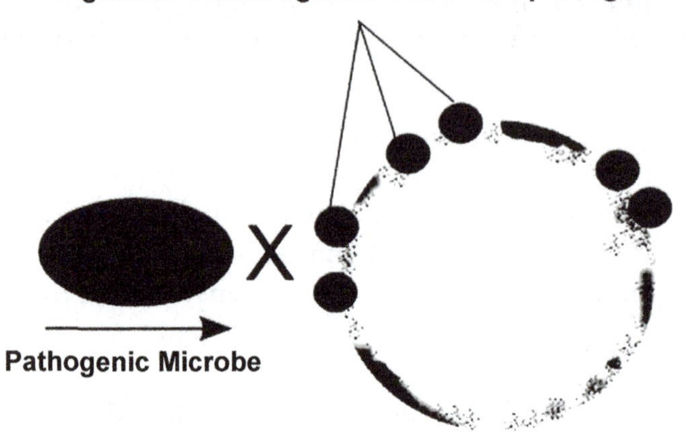

Our skin, which is our first line of defense, is a good example. Bacteria such as *Staphylococcus* and *Psuedomonas* inhabit our skin and prevent invasion and infection by seriously nasty species.[6] They do this thanks to their potent anti-bacterial and anti-fungal properties. As long as they stay outside our bodies, they are our allies. But these two genera themselves can cause serious problems if somehow they breach the skin barrier. In fact, *Psuedomonas* can be a problem in hospital acquired infections with a 50% morality rate in certain weakened patients.[7]

If you think about it, this amazing relationship challenges our everyday view of what is "us." Are these 300+ trillion microbes part of us? It's a difficult question to answer. If we could magically remove them from our bodies, we would lose our ability to adapt to changes in the environment and we would die. Looked at in terms of its protective function, *Psuedomonas* is actually acting as a branch of our immune system.

In our gut, the question of self vs. non-self becomes even less clear. Our gut flora contains about 30% of our microbial population (about 100 trillion). These microbes not only serve as an immune defense, but actually serve as an organ as well, an extension of our other organ systems and a necessary component for digestion. In other words, they do complex physiological functions that result in our survival.

An important function of our intestinal flora is to aid with carbohydrate breakdown. They take complex carbohydrates and convert them into a more simplified energy source that we can absorb. They also produce vitamins such as biotin and folate, and help with electrolyte transport. They even secrete factors to help stimulate growth of the intestinal skin lining (epithelial cells) after injury, and suppress pathogenic microbial growth. Literally, we could not survive without them.

Our gut flora also aid in the suppression of allergies and autoimmune conditions, such as Crohn's disease. They are even implicated in the development – or prevention – of obesity.

For all these reasons, it seems clear that our microbial friends serve as physiological extensions of ourselves. So again, we ask the question: What are we? Does the concept of "self" end with the epithelium – our skin? Or does it end with the microbes? The obvious answer is that the concept of "us" must make room for our microscopic friends.

There Is No Actual "Self" (sort of)...

Alfred Tauber has addressed this conundrum of self vs. non-self with respect to our mutual dependency with microorganisms: *Symbiosis challenges this well-entrenched definition of the individual organism, not only because physiological autonomy has been sacrificed, but anatomic borders have lost clear definition and development becomes intertwined among several phylogenetically defined entities.*"[8]

Friendly commensal microbes provide cellular feedback, immune protection, and synthesize nutrients

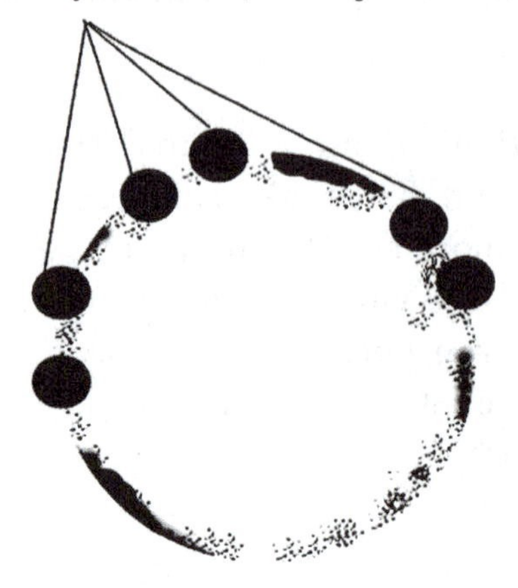

Tauber does a splendid job of making the case that there is no real "self" with a strictly demarcated boundary. In other words, he suggests that the *boundaries of the host organism are defined by immune activity.* That is, the line of distinction for an organism is due to the patterns of immune tolerance or reaction. Therefore, if the beasties in our gut do not initiate a defensive immune response, they are in fact "us".

The thrust of Tauber's ideas therefore is that **the boundaries of an organism are defined by immune reactivity**, and not by the organism's own tissue. This is a hugely compelling argument. However, Tauber only addresses self (vs. non-self) in terms of the immune system's reaction. So the fact that we have a symbiotic relationship with a host of microorganisms adds a layer of mystery to our concept of what is self and what is not. But there's more. Humans have a whole multitude of senses that input information into our system, and the microorganismal input (which is called chemoreception) is only one of many.

Our input mechanisms include:

- Sight (ophthalmoception)
- Hearing (audioception)
- Taste (gustaoception)

- Smell (olfacoception
- Touch (tactioception)
- Temperature (thermoception)
- Movement/space sense (proprioception)
- Pain (nociception)
- Balance (equilibrioception)
- Vibration (mechanoreception)
- Chemical states (chemoreceptors)

Technically, the immune system, which falls under the chemoreceptor category, accounts for only one input out of eleven when it comes to detecting what is 'us' vs. what is not. And to make things even more complicated, the immune system interacts extensively with the nervous and hormonal systems, which in turn, receive huge amounts of information from the other inputs listed above.

Despite the controversy, it does not take a rocket scientist to tell whether we have a "self" or not. As conscious humans, we intuitively know when our boundaries have been violated. For instance, when normally friendly intestinal bacteria take up residence in other places such as the urinary tract, or get ingested through contaminated food, we certainly know our self's boundaries have been crossed, because they make us ill. Equally, when we stub a toe or have surgery, we certainly become aware that 'we' have been hurt. When we are assaulted, our personal space has been violated. Perhaps akin to the friendly microbes being part of us, our home, our pets, our possessions can be psychic extensions of us.

When Does "Not Us" Become Part of Us?

A worthwhile question flowing from this: "When, after entering our system, does the stuff that is not originally part of us, such as food, water, or air, actually cease to be the non-self and become part of the self?" When does that oxygen molecule we breathe actually become part of us? Is it when it enters our nose, when it enters or lungs, when it actually penetrates our lung epithelium and binds with red blood cells, or when it actually gets incorporated into the target cell? The same can be said for the nutrients that we introduce into our digestive tract.

> ✓ ***Point to consider-When does matter or energy become part of "you" after entering your body?***

When we look at the central nervous system, the question becomes even more bizarre. When does the photon entering your eye's pupil, or the sound vibration hitting your eardrum, actually become *you*? Yes, these outside signals are transformed into nervous system friendly signals, which are then transformed into concepts with which we are familiar. But unlike nutrients, these outside energy signals really never become physically incorporated; it is more like we have interpretations of these energy sources, and so the self vs. non-self line of separation becomes even more confused.

Looking at ourselves from a big-picture perspective, we can conceptualize the human body as an undulating, ever-changing plasma-like structure. Our borders, however we define them, are flexible, providing more permeability to our outside ecosystem when we need it and snapping back into a more protective, rigid structure when we are not being forced to adapt.

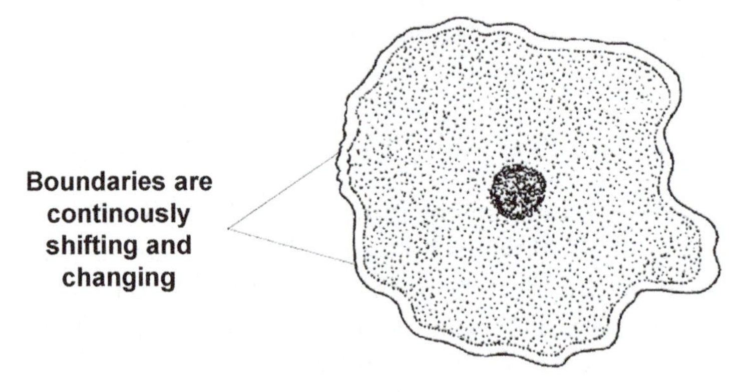

We are not so different from the humble amoeba

It is this plasticity that helps us adapt to the outside environment without going straight into organ failure and death.

How can we overcome these seemingly insurmountable obstacles for understanding the problem of self vs. non-self? Let's start by looking at the system as a whole, that is by taking a Gestalt (or holistic) approach and working backwards. But first, it might be useful to recap what we've established up to this point:

1. The human boundary does not end with the skin, or epithelium.
2. In the case of injury where the skin is breached, the self vs. non-self boundary starts to become a little more obvious, at least cognitively.
3. We incorporate knowledge from the outside world into our nervous system. How else could we do something like learn a new language, Spanish for example?
4. We take in atoms in the form of nutrients that are not us originally and make them us.
5. We take in photons, vibrations, pressures, etc., and transform them into permanent neural etchings.
6. We understand the significance of self vs. non-self in physical assault or injury.
7. We certainly understand self vs. non-self when talking with other people. They have their ideas and physical boundaries, and we have our own.
8. We extend our sense of self to our possessions, such as our home, clothing, cars and sentimental keepsakes

This new way of looking at ourselves clarifies the very disorganized (if not inaccurate) understanding of our current medical consensus, as it solves many mysteries regarding the way an organism adapts to its environment. It is highly relevant to the questions we are asking about stress, because the stress response relaxes the limits of what we think of as "self."

Key Concepts:

- Identify and define the following concepts:

  Agency
  Free will
  Complex adaptive systems
  Gestalt
  Reductionism

- Identify and name examples of each of the complex adaptive system's seven attributes.
- Explain the significance of Alfred Tauber's and Kurt Goldstein's work.
- How many cells does a typical human have? This includes ourselves as well as "non us" flora dwelling on and inside our bodies.
- What would happen to our bodies if suddenly we lost all our flora?
- What are some of the functions of our flora?
- A what point does the boundary between "self" and "non-self" begin and end?
- Identify the eleven forms of human sensory input.
- According to Goldstein, what happens when we try to study an isolated part of a complex system without regard to its relationship to the whole?

CHAPTER THREE

HOW DOES THE STRESS RESPONSE HELP US?

Chapter objectives

- ✓ ***Without the stress response, we would die very quickly.***
- ✓ ***When the stress response mechanism becomes fatigued, the affected body part becomes exhausted.***
- ✓ ***Through the stress response, we can change our physical bodies to adapt to the environment.***

What We've Learned So Far...

Let's take stock. We know now that our body has two states - one rigid and defensive and the other open and adaptive. We alternate between these two states many times, not just during our lifespan, but also within each day. The two states are, incidentally, ancient organismal responses that evolved side by side. Switching back and forth between them forms the basis of the stress response mechanism.

If you had been created without a stress response, it is unlikely you would have survived the massive adjustment of birth. A complex creature that is unable to summon up a stress response when needed would simply go into systemic shock and die from cardiac failure.

> ✓ *Without the stress response, we would die very quickly.*

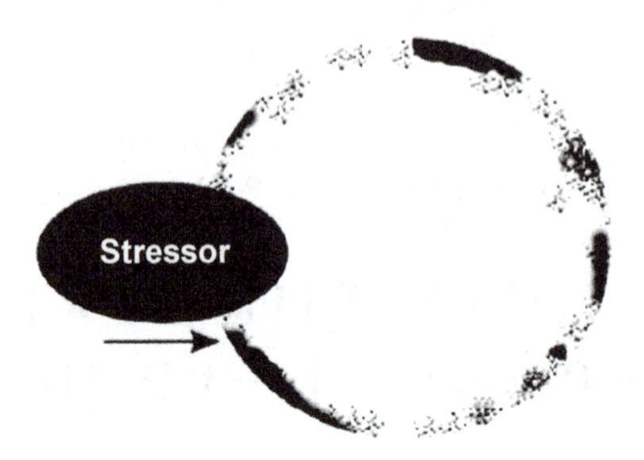

When organism forms a plastic state, the elements are incorporated into its morphology via a "stressor"

Another way of looking at this is to say that when an organism's defenses become too strong, and its borders to the outside close too tightly, the organism will fail to adapt to its environment. If this failure continues, the tissues, organs, and eventually the whole organism will become exhausted, and its functioning will become compromised, and unable to respond. If the exhaustive state continues, ultimately the organism will die.

What do we mean by "exhaustion"? We will be examining exhaustion in detail later in this book, because it is so important. For now, a brief explanation will suffice. Exhaustion in this context is Selye's description of *organ (or tissue) dysfunction, or even death, due to persistently high levels of fight-or-flight activity, as well as the consequences of high exposure to stress-related hormone or immune by-products.* This happens when our reserves are used up, and we no longer have the resources to continue adapting to pressures in our environment. We then become overwhelmed.

> ✓ ***When the stress response mechanism becomes fatigued, the affected body part becomes exhausted.***

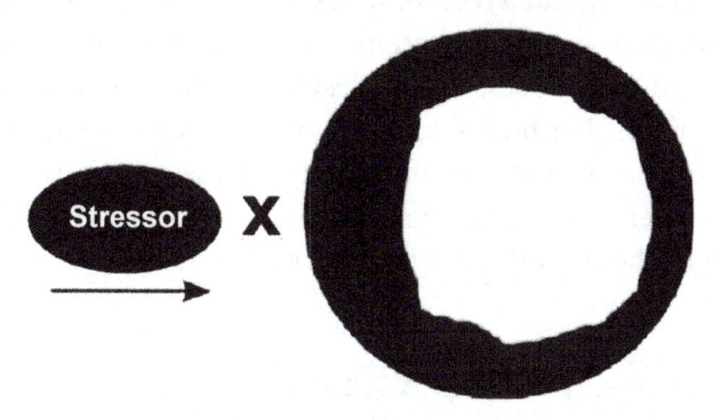

The organism forms an inelastic and more defensive posture. This make the organism more relatively resistant to the environment

The way this happens is that, as part of our body's defense mechanism, when the stress response kicks in, fluid transfers from inside your plumbing (the blood vessels) to the extra-cellular space (a process involved with inflammation), causing both nutrient and waste transfer to slow down drastically. In the short term, the cells can cope. If it continues for too long, it causes cells to die from both starvation and exposure to their own toxic excrement.

Putting Our Stress Response into Context

How does the stress response work in real life? There are literally hundreds of stress triggers (or *stressors.*) However, as the field of stress study has evolved over the last hundred-plus years, we have categorized them into three major classifications. They are as follows:

1. **Physical stressors:** Any adjustment you have to make to cope with mechanical stressors, such as exposure to the cold, heavy weights, manual labor, or tight shoes.

2. **Chemical stressors:** Adjustment needed to cope with exposure to drugs or poisons, to excessive food or poor quality food, and to internal states, like blood glucose or pH swings.
3. **Psychological stressors:** As animals begin to have more complex nervous systems, adjustments due to excessive cognitive processing or restructuring become important stressors. For humans, having the burden of such a freakishly large brain cortex is itself a huge stressor. Robert Salposky, a pioneer in modern stress research, discusses this thoroughly in his book "Why Zebras Don't Get Ulcers."

Routine vs. Novel Aspects of Reality

We have seen that to cope with these stressors, we tend to switch between an open, adaptable state and a closed, defensive state. This dual system is not the only duality relevant to our story. There are also two basic patterns of nature for which we have evolved over millions of years. One pattern involves the repetitive, more **routine** aspects of nature, and the other involves the interruptive, **novel** aspects. As we will find out shortly, the routine and novel aspects of reality are integrated into our being via the machinations of the *autonomic nervous system.*

Some examples of the repetitive elements of nature include the day/night cycle, seasonal cycles, reproductive cycles, feeding times, and so on. There is a cyclical nature to these patterns. Organisms have evolved over millions of years in anticipation of these cycles, and without these fundamental repeating patterns in nature, life is simply not possible.

In contrast to these repetitive patterns, we can also have major disruptions in our lives that jeopardize our very survival. When someone close to us dies, when we lose our income or job, when we move residence, when we (or someone close to us) get seriously ill, or when we get divorced – these are major shocks to our existence.

We become very aware of the major reorganization and adjustments needed to survive these major loses. Our grandmothers intuitively knew that exposures to these sorts of shocks could destroy a person's mental or physical health, or even kill them.

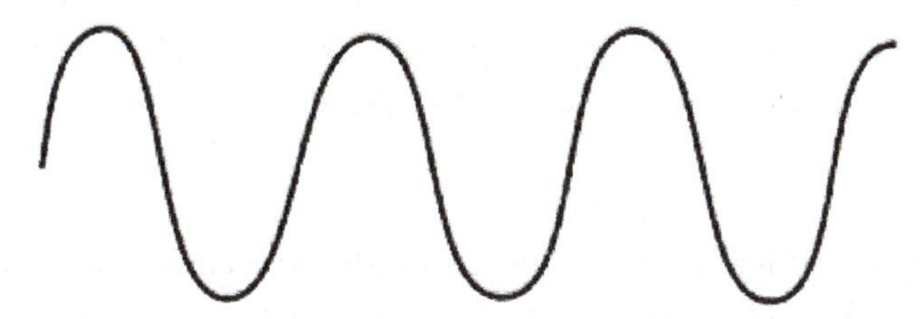

One state of nature has repetitive (or cyclical) elements. These include predictable routines whose behavior can by accurately anticipated in time and space.

Why? It all comes down to energy conservation. The living process uses a tremendous amount of energy. Mother Nature does not waste energy, as it tends to be in short supply (like money) for most of an organism's life. Therefore, if we are not presented with a stable, "basic" foundation in life, we simply fail to thrive. Here in Ecuador, where I am writing this book, the rain forests need warmth and a steady supply of rain. An unimaginably complex network of plants and animals has developed in accordance with these unique ecosystem rhythms.

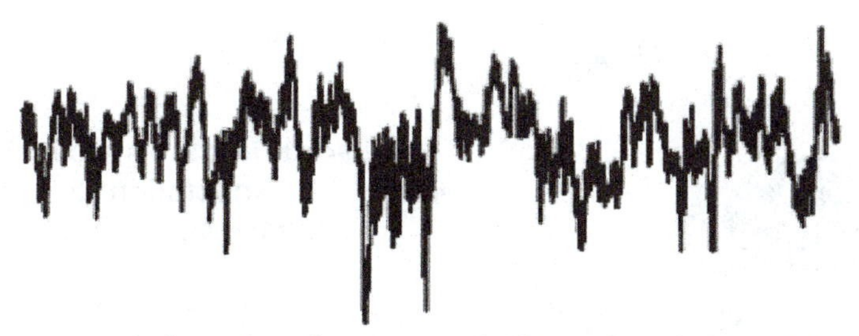

When events in nature become volatile and unpredictable, the physiological stress response is required to keep organism in composure to adapt to novel elements.

If we think about a human baby, the importance of rhythm to the human development process becomes crystal-clear. Many aspects of a newborn are de-synchronized. Consider the infant's sleep/wake times, feeding demand, and even gut bacteria. If an infant does not have access to repetitive input, many components of the nervous, digestive, immune, cardiovascular, and nervous system will not develop. A human being must have some sort of "reliable" aspect to her environment in order to survive. We must all have repetitive inputs that we can rely on, so that we can develop as healthy human beings.

However, as cyclical as life is, we know all too well that there are also "shocks" in life. These shocks come in all sizes and flavors. Examples of small shocks include skipping or changing the composition of a regular meal, or going on a vacation to places of differing climates, altitudes, or time zones. Another example would be going without our usual amount of sleep or changing our work schedules for a few days. You get the picture.

Shock and Survival

Sometimes there are major disruptions in life that jeopardize our survival. People who are fortunate to have a healthy, well-functioning stress response, and some good luck, will survive these shocks. And if all goes well, they may even reach a new level of personal growth.

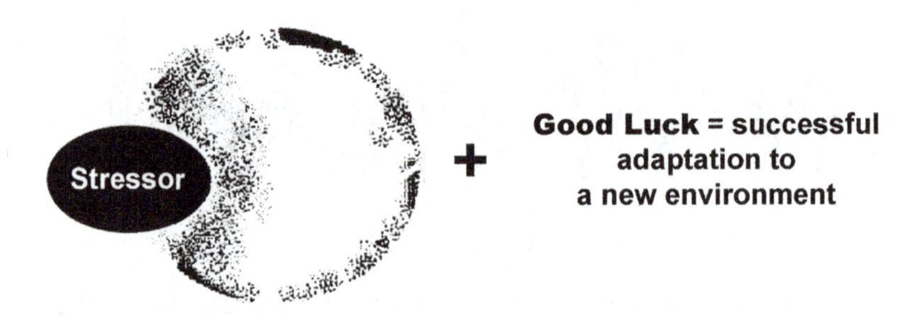

This idea of being able to survive shocks is a pivotal concept to truly understanding the stress response, so let's look at what it means on a deeper level.

It may help to play out a scenario. Let's say you lose your job. You now have no income but many bills. In order to adjust to this stressor, you must undergo a massive amount of reorganization. You switch from a perspective of abundance to that of scarcity. You stop eating steak and drinking lattes. You worry about rent, and along with worrying, you may isolate yourself from family and friends with trying to deal with the overwhelming feeling associated with insecurity. Your entire world has turned upside-down. The process becomes very painful.

Humans are very special creatures, in that our large cerebral cortex enables us to plan and control our environment, to an extent. This has the potential to reduce the strain (or lessen the metabolic cost) of having to *change ourselves physically to fit the environment.* By manipulating the environment, we can lessen the body's need to adjust to environmental stressors. We could not live in a freezing climate in our natural state, but we are able to manipulate the environment by sewing clothing, constructing shelter, and creating warm fires to make living in such conditions possible. We can do these things, thanks to the foresight provided by the prefrontal cortex.

> ✓ ***Through the stress response, we can change our physical bodies to adapt to the environment.***

Going back to your lost job, and you will realize there is a limit to how much influence you can exercise. Much like a loved one dying, there are just some life events where the influence of the environment eclipses our own. The prefrontal cortex cannot plan, scheme, or connive its way out of many of these overwhelming environmental demands, so we find that we are forced to adjust.

To help us cope with these situations, nature has provided us with a stress response. We have seen that the healthy set stress reflexes serve to change out biological structure to a more plastic, flexible state. This enables us to alter *ourselves*, both mentally and physically, adapting our biology to the needs of the environment.

How Humans Reorganize their Shape during Stress

Small adjustments are of small consequence, so the distortion of one's body shape (or *morphology*) is routine and minimal in these instances. However, the unemployment example above requires that we change ourselves significantly. We go through a grieving process for the loss and for the psychic injury incurred by realizing that there are things in life that we just can't change. We modify our spending and living habits. Hopefully, we will go into a "hunt" mode to try to procure new resources.

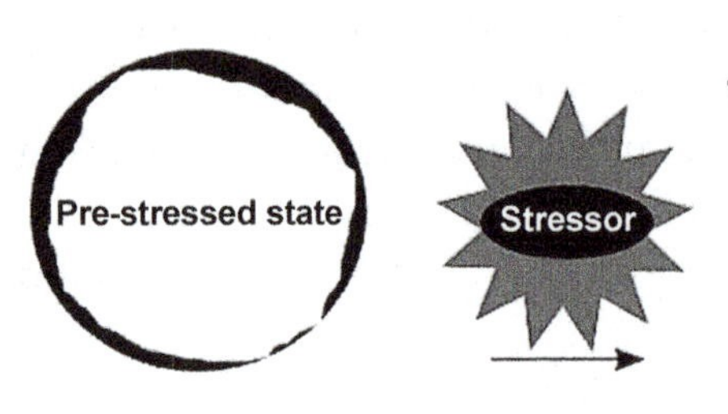

CHOICES WE HAVE

1. Can I change the environment to lessen my stress?

OR

2. Do I need to change myself in order to adapt to the environment (because the environment is uncontrollable)?

The point here is that there are times in our lives that we have to reorganize ourselves, because attempting to reorganize the environment has little, if any, effect. As a matter of fact, trying to change the environment when we need to change ourselves reduces our "fitness," and makes us even more vulnerable to the ravages of an unkind environment.

We probably notice this phenomenon most within our thoughts and emotions. Our thinking and feeling apparatus are often deeply rooted. Reorganizing our minds takes a lot of work. Entrenched neural networks, especially in our more ancient emotional centers, must now adapt to the new situation. This is why the process of grieving or adjusting after any type of loss is a very important step to go through, and not just with the death of a loved one.

Since we are on the subject of psychological adjustments, we must address the times that we fail to adapt ourselves to new realities. This is the core issue that results in mental illness.

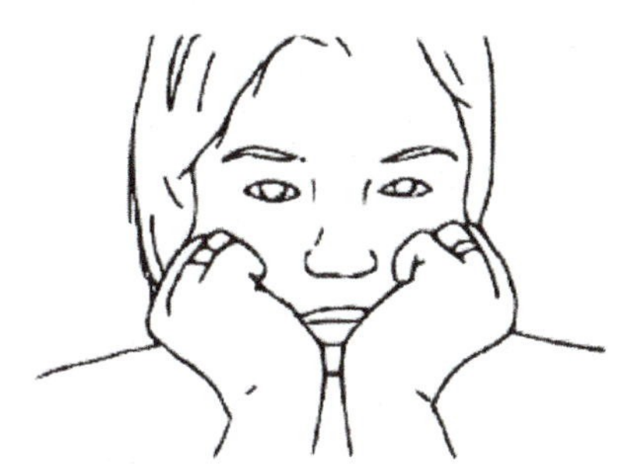

Problems with processing stressful situations are the roots of all mental illness

Perhaps we employ denial, or even more common, dissociation. Perhaps we get stuck in pathological grieving. For whatever the reason, we fail to acknowledge and adapt to our new personal reality. We will look at this more.

When we view an organism's response to the environment in this new light of novel vs. cyclical, many misunderstood aspects of biology become clear. Most of our life is spent doing routine things—sleeping, eating, working, etc. We came to survive as organisms because we could trust that certain aspects of our environment were reliable. Think about this: if we did not have predictable growing seasons, would agriculture be possible? Similarly, if your clansmen or family are unpredictable and unreliable, then how can you possibly survive as a child?

Key Concepts:

- What is Selye's definition of exhaustion?
- Identify the two states of reality presents to us. What types of events occurs in each of these states?
- Identify major "shocks to existence" you have experienced in your life.

- How can our brain protect us against stressors?
- What two choices does an organism have when confronted with a demand for adaptation?
- What happens when an organism does not have stable, repetitive input?

CHAPTER FOUR

THE ADAPTIVE NERVOUS SYSTEM

Chapter Objectives

- ✓ ***The autonomic nervous system provides automatic control of the body. It regulates tissues to be more opened or closed to external flow of mass, energy, and information.***
- ✓ ***The parasympathetic branch processes routine environmental elements.***
- ✓ ***The sympathetic branch processes novel environmental elements.***
- ✓ ***The stress response ultimately starts at the individual cellular level.***
- ✓ ***Stress proteins are like having intelligent butlers in your cell. Whatever the type of mess results from stress is repaired, resorted, refolded, or cleaned up.***

Understanding the Autonomic Nervous System

Our nervous system is constructed to process and respond to both the repetitive and the novel aspects of the environment. We even have a mechanism called the autonomic nervous system which functions as a physiological gate keeper, regulating the flow of mass, energy, and information across our body's borders.

> ✓ ***The autonomic nervous system provides automatic control of the body. It regulates tissues to be more opened or closed to external flow of mass, energy, and information.***

For the most part it is automatic, requiring little to no conscious effort, and it controls the basic vegetative functions of pretty much every one of our organ systems. As mentioned earlier, this part of the nervous system filters, disentangles, sorts, and then processes both the novel and the routine parts the environment presents to us.

The autonomic nervous system contains two primary branches: the *parasympathetic* branch, which regulates responses to routines aspects of life, and the *sympathetic* branch, which processes and helps us incorporate and respond to novel events. For instance, the sympathetic nervous system controls the "fight-or-flight" adrenaline based system that responds to novel demands by making the body more open to the external environment.

> ✓ ***The parasympathetic branch processes routine environmental elements.***

Functions of the parasympathetic nervous response:

- Constricts pupils
- Increases salivation
- Decreases breathing rate
- Decreases heart rate
- Dilates blood vessels
- Stimulates digestion
- Contracts bladder muscles
- Stimulates bowel movement

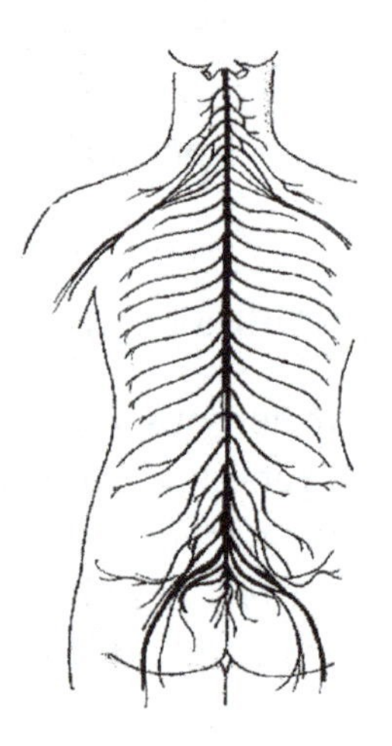

- Promotes sexual arousal
- Anti-inflammatory and anti-shock effects

> ✓ ***The sympathetic branch processes novel environmental elements.***

Functions of the sympathetic nervous response:

- Dilates pupils
- Decreases salivation
- Increases breathing rate
- Increases heart rate
- Constricts blood vessels
- Inhibits digestion
- Relaxes bladder muscles
- Inhibits bowel movement
- Promotes orgasm

How does an internal stimulant, such as adrenaline, transform human form into a more plastic, malleable state? If we look at the diagram above, we see that one function of the sympathetic nervous system is to dilate the pupils. When our pupils dilate, they allow more light (in the form of photons) to enter the eye. This allows the eyes to become more permeable to light. As a result, more visual information enters the nervous system.

Another example is the relaxed bronchial airway system. In this case, the airways dilate, allowing more oxygen to pass though, diffuse across a capillary membrane, and attach itself to the hemoglobin in red blood cells. So we again become more open to letting in oxygen.

Interestingly, we find that the sympathetic branch controls human orgasm. This makes sense, as sexual activity by its very nature "crosses the line" of personal boundaries. An attempt to fertilize an egg is the ultimate example of one person's boundary crossing into that of another.

All the sympathetic nervous system's effects on the digestive system serve simply to shut functions down during times of adjustment. If an organism is becoming more plastic so that it can rearrange its body to fit the environment, constraints of nature will not allow for active repair and feeding of the tissues. Regarding higher cognition, it is tightly involved with processing of novelty and increased engagement with the environment, presumably due to stress hormone activity.[9] That is, it enables your thinking to be more "open" to the effects of the environment.

The effects of the sympathetic nervous effects on the immune system and resultant inflammation are very interesting. The sympathetic nervous system has been shown to have potent anti-inflammatory responses, thus lowering defenses so that adaptation to any demanding stressor can take place.[10]

To be fed and repaired, the tissues must be under the influence of the evolutionarily even more ancient *parasympathetic* response. That is, the body must assume a more rigid, defensive stance that makes it relatively more resistant to the effects of the environment.[11] Cognitively, stimulation of the major parasympathetic nerve impairs creativity and flexible thinking and well as promotes more withdrawal from the environment.[12]

Both your bladder and intestines are normally relaxed, thus keeping foreign, non-self material (urine and feces) retained. However, with the "rest and digest" parasympathetic response, your nervous system recognizes non-self (when you need to *go!*) and gives you the sensation of needing to use the bathroom.

Perhaps one of the strongest illustrations of self vs. non-self distinction is when you have to hold your bladder for much longer than usual. As the bladder spasms, we are strongly reminded of the need to eliminate the non-self material from us.

We can also see the effect of parasympathetic activity on digestion. Again, its function is all about maintaining the structural aspect of the self. When you eat food that is not agreeable, your parasympathetic tags it as 'non-self' and induces the vomiting reflex. Much like needing to use the bathroom urgently, throwing up orients the entire organism to rid itself of non-self material. It also is activated when your stomach is distended, indicating fullness from food that is waiting to be digested, as technically it is not part of you *yet.*

The Stress Response on a Cellular Level

The autonomic nervous system helps us modulate the stress response on a whole body level. To understand the stress response in more detail, we need to look at its effect on a cellular level. The first part of a cell that experiences stress is the fat-based (or lipid) cellular membrane. Picture this (as shown below) as a very thin, supple layer, and you will understand why it is affected by chemical and mechanical stresses, like temperature, physical impact, pH, salt concentrations, toxins, signaling hormones, etc.

> ✓ ***The stress response ultimately starts at the individual cellular level.***

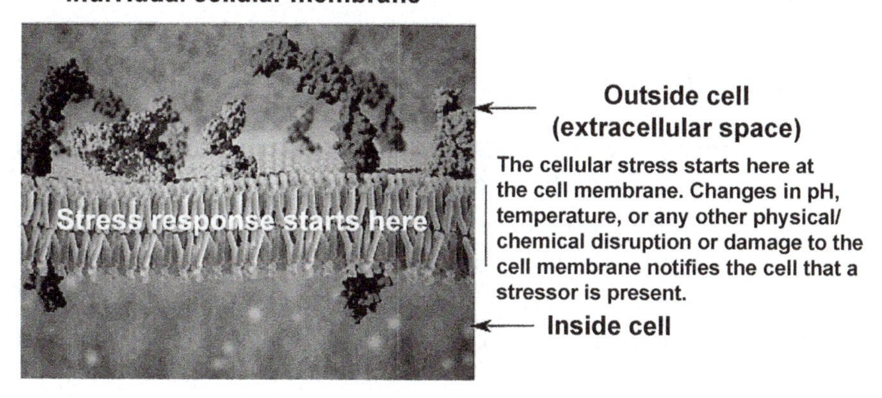

After a stressor stresses the cellular membrane, specialized heat shock proteins (proteins released during stress) become activated

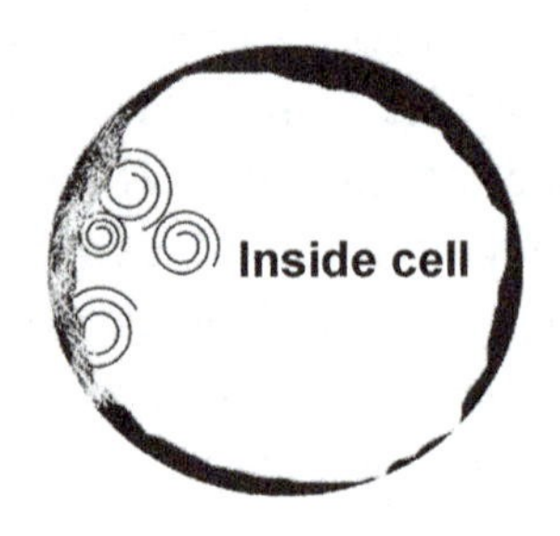

When this happens, a cascade of events occurs to aggressively activate early stress detectors. One specialized mechanism that is essential for elasticity on the cellular level is called the heat shock protein (HSP). Though discovered initially during severe heat stress, it has long been established as one of the major catalysts for cellular adaption.[13]

There are many cellular stress responses that are activated when stressful events happen. They include[14]:

- **The Unfolded Protein Response:** Works similarly to HSPs, but in a different part of the cell.
- **DNA Damage Response:** Fixes breakages in DNA due to physical and chemical stressors (e.g. UV light).
- **Oxidative Stress Response:** The use of oxygen in metabolic process produces a lot of reactive waste (lots of restless electrons eager to react). This process clears out this type of waste.

The HSP Response

The heat shock protein response is among the most important stress responses, and for simplicity's sake, we will focus on just this one. HSP is a fascinating protein, as it is very similar across many species of plants and animals.[15] This should not be surprising; the overall stress response shares many similarities across many species of plants and animals. Like basic functions, such as DNA replication and protein synthesis, the functioning of HSPs is also remarkably similar in many cell types across all species.

HSPs' job is to clean up the mess created by core life functions. Under normal non-stressed conditions they make up to 5-10% of a cell's protein content, but when the stress response gets activated, this can go up to 15%. For just one protein family controlling just part of the cellular stress response, occupying 5-15% percent of total protein volume is a massive undertaking. What this tells us is that a large part of our cellular machinery is dedicated just for adapting and adjusting to the environment, and not just for growth, repair, or basic housekeeping functions.

This reminds us that cells do a lot more than just the basic functions of eating, working, and releasing waste. A significant part of the cell's structure and functioning is designed for adapting to changes in the environment.

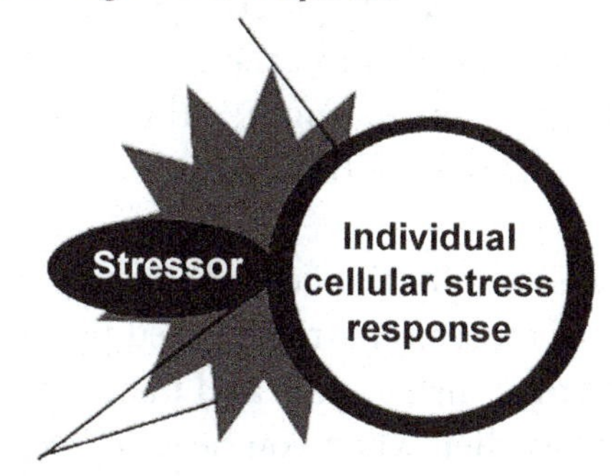

Acting as the first line of stress signaling for the cell, HSPs are fascinating critters. The smaller members of the protein family are activated when the cell membrane is heated, sheared, or made more liquid through chemical insults.[17] If the stressor gets really intense and starts to cause damage to the DNA, cellular organs, or other proteins, larger HSP members come in and carry out all kinds of intelligent intervention.

> ✓ ***Stress proteins are like having intelligent butlers in your cell. Whatever the type of mess results from stress is repaired, resorted, refolded, or cleaned up.***

Damage to the cell can be actual physical damage from the stressing event (a physical or chemical impact) or due to the strain caused by increased cellular metabolic activity. So the impact of the stress can come from both external and internal events.

When a potentially fatal stressor impacts the cells, causing severe damage to the molecules, HSP and other responses kick in to start repairing. Protein repair and function are very important to the overall functioning of the cell, and it is proteins that do the work needed to keep an organism alive and functioning. Think of HSPs as mini engines that combine together like legos, carrying out a variety of functions.

When a stressor hits the cell, HSPs and other factors ensure that production of most non-stress related proteins comes to a standstill. Growth, repair, construction, and maintenance pretty much halt. As we discussed earlier, when you need to become more elastic to meet the needs of the environment, it is counterproductive to put energy into growth and repair. Some processes of growth and repair can themselves be stressful, because of the sludge buildup and build up of metabolic mistakes over time with metabolic activity.

In fact, if the cellular machinery needed for growth were to be active during the stress response, the molecules synthesized could overwhelm the protective cellular stress response. It is much like eating a huge Christmas dinner, and deciding to swim the English

Channel immediately afterward; such an undertaking would be likely to inhibit your functionality!

Now that other functions have stopped and the HSP family of proteins have a clear run, they go in and fold unfolded proteins, fix damaged ones, and eliminate those that are so messed up that they are literally the molecular consistency of scrambled eggs. They also allow for stress tolerance, which we see with athletic conditioning, and tolerance to heat and other events that build adaptive resistance.

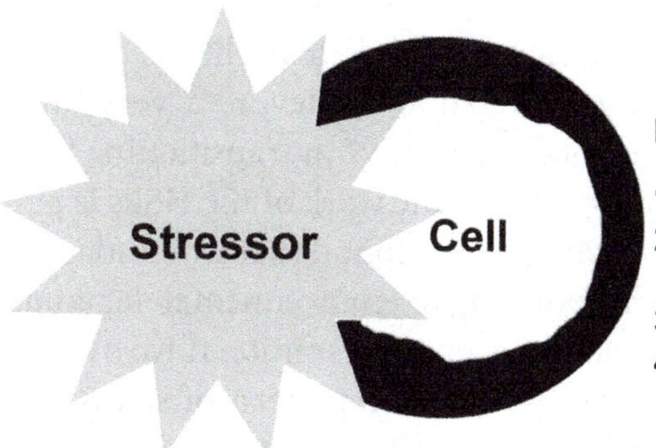

Impact of stress on a cell:

1. **Damages protein**
2. **Unfolds normally folded proteins**
3. **Damages DNA**
4. **Damages cell membrane**

Heat shock proteins provide several different functions within the cell, including[18]:

Individual cell undergoing stress

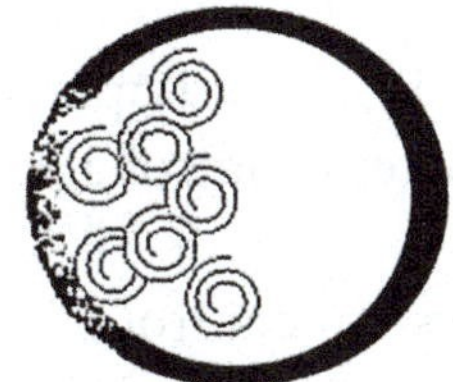

An intense physiological reaction involving HSP occurs during the stress response.

This results in less:
Growth
Repair
Reproduction

- Correcting the folding of newly constructed and stress-accumulated misfolded proteins
- Allowing the cells to survive lethal conditions (HSPs inside the cell)
- Directly interacting with various components of programmed cell death machinery (HSPs inside the cell)

- Mediating immune functions (HSPs outside the cell)
- Stabilizing new proteins to ensure correct folding
- Monitoring the cell's other basic functions (involving protein maintenance)

The major role played by HSP proteins brings us to a very interesting conclusion. A huge volume of protein of our body's infrastructure is reserved for adaptation. In other words, there is much more to life than basic growth, maintenance, and repair. A big part of our biological machinery is dedicated to cleaning up the mess created by having to adapt to changes in our environment.

You can see now how important biological flexibility is. It is so important, that our bodies are prepared to devote a very large amount of what Selye calls "adaptation energy" in transitioning from the defensive state to the adaptive state. The work of the HSPs is part of this adaption energy. Importantly, Selye found that while adaptive energy allows for seamlessly responding to environmental demands, the available quantity of this adaptive energy is finite. If we overuse it, problems arise. Selye describes a unique pattern of stress that follows, and which he called the **General Adaptation Syndrome**. We will discuss this in more detail in the next chapter.

Key Concepts:

- Define "autonomic nervous system". Identify and define its two branches.
- What is the "fight-or-flight" response? Why is it important?
- How does the sympathetic nervous system control allow an organism to become more "open"?
- How does parasympathetic nervous system control allow an organism to become more "closed"?
- Which part of your cell detects a stressor? What types of stressors can it detect?
- Identify and explain the purpose of the four major cellular stress responses discussed.

- Focusing solely on heat shock proteins (not including other stress-related proteins), what percentage of a cell's protein volume does the HSP family make up in both stressed and non-stressed states?
- What are some important functions of heat shock proteins?
- Define "adaptation energy".

CHAPTER FIVE

SELYE'S GENERAL ADAPTATION SYNDROME

Chapter Objectives

- ✓ ***The General Adaptation Syndrome is a three stage process that permits an organism to adapt to a stressor.***
- ✓ ***The Alarm Stage has two parts. The first part involves an initial tissue confusion or "shock" and the second part involves starting the adaptive response.***
- ✓ ***The Resistance Stage inhibits the shock in the Alarm Stage. In this stage, the tissues reorient and become highly responsive to the demands of the environment. This is adaptation.***
- ✓ ***It is critically important not to add extra stressors to the stressed tissue during the Resistance Stage.***
- ✓ ***If the stressor persists and is not resolved, tissue and cellular damage occur. The ability to adapt lessens.***
- ✓ ***If the Exhaustion Stage persists, then irreversible tissue damage or death occurs. The stressed area resumes the inflammatory process and adaptation becomes difficult, if not impossible.***

The Three Phases of Adaptation

Hans Selye made many brilliant observations about the biological stress response over his long career. He described an organism's pattern of adaptation that included three phases: Alarm, Resistance, and Exhaustion.[19]

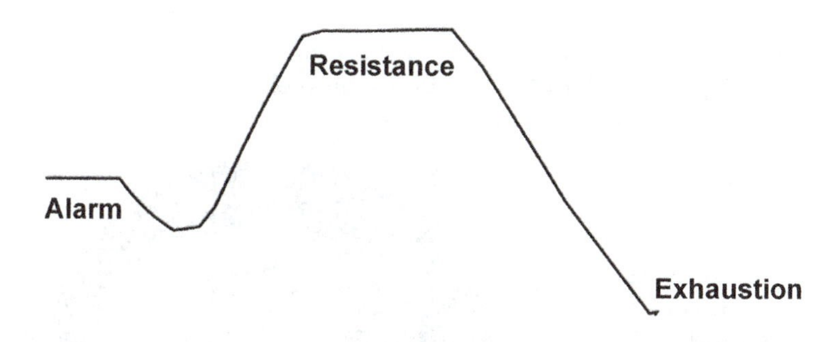

> ✓ ***The General Adaptation Syndrome is a three stage process that permits an organism to adapt to a stressor.***

The Alarm Phase

Picture this: You are peacefully driving your sedan on an unobstructed road thinking about lunch. Your parasympathetic system is running things smoothly until you come around a blind corner and discover that you are heading straight for a ten-car pileup.

We all know what happens next. Your blood pressure and heart rate go through the roof. Your consciousness becomes very focused. Your pupils dilate, thus retrieving more visual information about the environment. In an instant, you become fully aware, vigilant, and ready to make fast driving moves. Unfortunately, it is too late, and you crash into the pileup.

The alarm phase has already kicked in. You are not in the homeostatic phase as you were a moment ago. You start to go into *shock*. The experience of shock is an attribute of the parasympathetic system, and when it occurs the contents of the blood vessels leak out into the extra-cellular space, also known as *interstitial space* (see figure below).

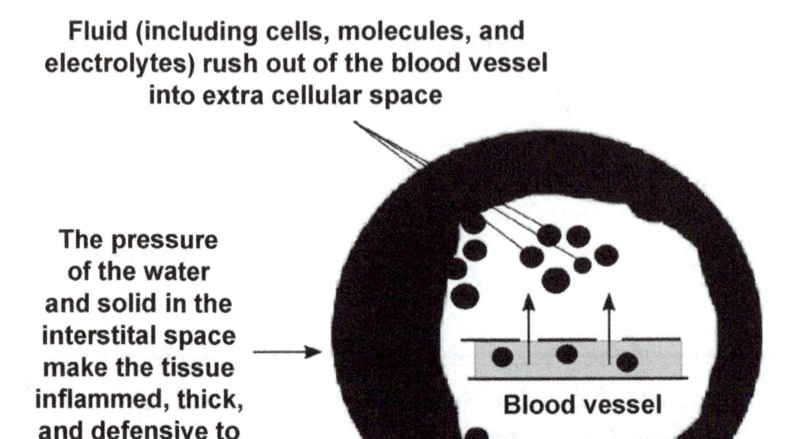

When the body fluid mixture leaks (more like rushes) from inside the blood vessel into the extra cellular space of the stressed or injured area, the area builds up pressure and becomes warm and *hard to the touch.*

> ✓ ***The Alarm Stage has two parts. The first part involves an initial tissue confusion or "shock" and the second part involves starting the adaptive response.***

Shock is the extreme systemic form of inflammation, except that instead of being localized to a specific region (like a twisted ankle), the swelling is happening all over your body. Your blood circulation slows, as it is collecting outside the tissues to immobilize the stressed area and to deliver defensive and cleanup measures. Blood pressure drops, and with it the force of the extra-cellular fluid to act as a delivery and removal agent becomes far less efficient. The tissues cannot get their nutrients delivered or their waste carted away quickly.

In our road accident scenario, the accident has resulted in an injured leg. Inflammation begins, but the injury is not so great as to go into total shock. The action becomes localized to the injured area.

We have already looked at what happens next. Because the tissue is painful and hard to the touch, integration with the environment shuts down. The area loses its flexibility, and thus its ability to adapt. At the same time, at the cellular level, a huge avalanche of HSPs get produced. As we have seen, cellular maintenance and repair come to a standstill.

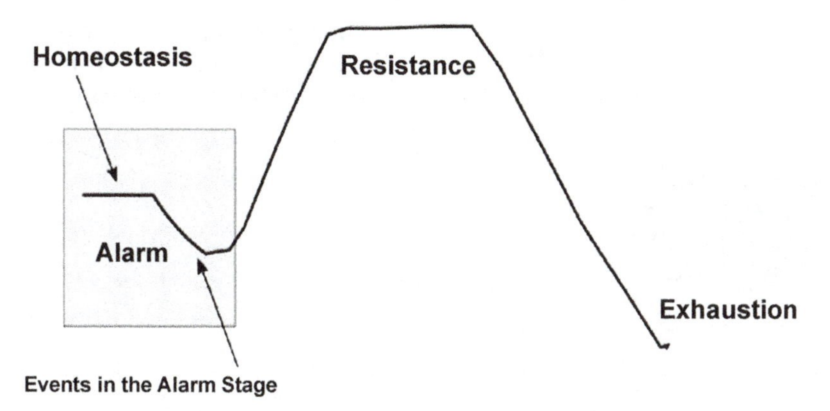

Events in the Alarm Stage

1. Individual cellular membranes affected by stressor
2. HSP proteins become active
3. More HSP recuitment because of stressor leads to future damage
4. Inflammation ensured to keep affected area mobilized
5. HSPs, as well as other stress proteins, force a hard "reset" in the affected tissue. Memory or body organization is temporarity erased

Put in simple terms, **inflammation induced immobilization + cleanup at the immune system levels (big chunks of debris) and at the cellular levels (damaged molecules with mishaps) + protein repair and reorganization = temporary biological reset.**

What exactly is being reset? It is the shape and function of the tissue that underwent the stress. It is an attempt to reorganize either back to its original state, or to a new, more adaptive state, and represents a profound level of the body's morphological memory. The nervous, hormonal, and immune mediators of inflammation shock are much older evolutionarily than our fight-or-flight sympathetic response. Shock stretches way back into our evolutionary history, and despite being an extreme reaction it is the default response to sudden changes. In fact, it is the very ancient subsystem branch of the parasympathetic system that controls many aspects of shock.[20]

Many changes occur, and most of them are beyond the scope of this book. But the most important point is that in this first part of the alarm response, the parasympathetic system gets confused, and the affected cells start losing memory of your body's normal form and structure.

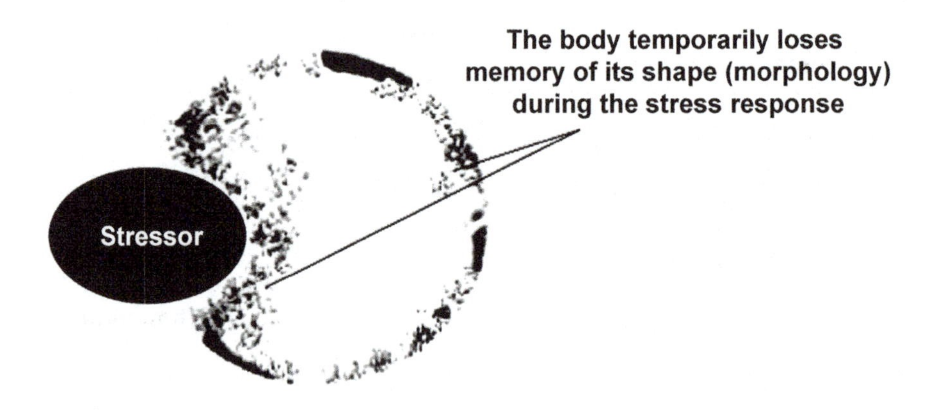

Remember, the parasympathetic nervous system's job is to deal with the major part of our reality that contends with cyclical/routine events. You literally go into dying mode when you go into shock, as this is how we all die. All death, no matter what the cause, is the result of a shock response that has lost control.

Inflammation of an injured body part is like systemic shock on a smaller scale. The local tissue starts going into a non-adaptive, non-functional mode. The tissue literally starts to become undone, as the earlier molecular memory of its physical infrastructure undergoes a hard reset. It is not appropriate to be defensive at this moment. It is appropriate at this point to be adaptive and flexible in responding to the environment, and thus the demand for the tissue to become "temporarily" more plastic.

For example, in the latter part of the alarm phase, our sympathetic branch kicks in, along with a strong recruitment of HSPs at the cellular level. In fact, epinephrine, a major fight-or-flight hormone induces the activity of HSPs in the laboratory. This could be a link that ties the stress activity within the cell directly to that of autonomic nervous system.[21-22]

Going back to the car accident example, your body releases a stimulating chemical norepinephrine (via the nervous system), a slightly slower epinephrine (released as a hormone in the blood), and the even more slowly acting (but longer lasting) steroid hormone cortisol.

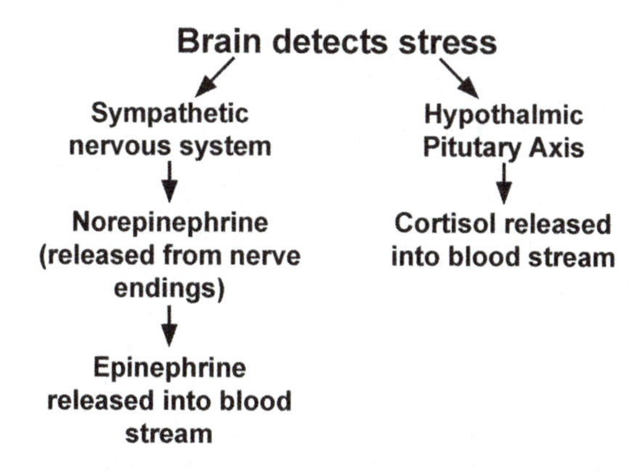

You start to pull yourself together within a few seconds to a few minutes. You may have been unconscious for a few seconds, but you are not hurt bad enough to stay that way. You regain consciousness, your blood pressure increases. You have no pain at the moment, but you realize you have an injured leg. You are messed up mentally and physically, but at least you are conscious and a biological memory of self.

The Resistance Stage

This stage can come on and resolve quite quickly, or it can last for many months or even years, depending on the severity of the stressor. The resistance stage is the stage that confirms that you are successfully adjusting to stress by adapting to the challenging situation. Cortisol, adrenaline, a variety of immune elements, as well as a cascade of events at the cellular level have adjusted the composition of your body to become more flexible.

> ✓ ***The Resistance Stage inhibits the shock in the Alarm Stage. In this stage, the tissues reorient and become highly responsive to the demands of the environment. This is adaptation.***

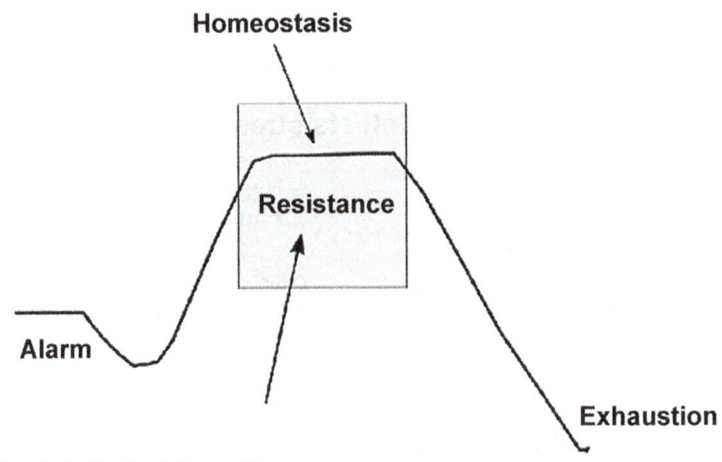

Events in the Resistance Stage

1. If successful, then new level of homeostasis is achieved
2. Cortisol is a slower and longer lasting hormone that inhibits most inflammatory activity. HSP assists with activating cortisol via the DNA level
3. Inflammation is inhibited and much of the stressor's cellular damage has been neutralized and cleaned up be HSP
4. The tissue affected reorganizes itself into a new interaction with the environment
5. Now that the tissue is "primed" by a stressful event, it is now vulnerable to additional shocks

Returning to our car accident example, by now you are hopefully conscious and can use the physical effort to try to get out of the car and make a phone call. You may even have the superhuman strength needed to drag yourself away, despite having a broken leg. You can thank your robust stress response mechanisms for keeping you alive through the trauma.

What exactly is going on at this point? The affected tissue has now had enough time to respond to the effects of the cellular stress response. The HSPs have started reorganizing the disorganized internal environment of your cells, and if the tissue has been stabilized, a high cortisol level inhibits the factors that promote inflammation. The tissue is being reorganized to "fit" the new reality, and is thus adapting. Depending on the type and severity of the stressor, this phase begins from minutes to days after the exposure.

However, there is a caveat - a huge one. Understand that at this point, you have taken out a huge biological emergency credit card loan with 40% loan shark interest and henchmen ready to collect on the balance. As time progresses (the length of which is related to the intensity of the stressor), there is a limited amount of "adaptation energy" available for adjusting to the environmental demands.

At this point, it is very important not to add to the burden on the tissue, by piling on more stressors. This is what successfully adapting to an environment is all about. The inflammation and pain experienced in the alarm phase serve to arrest movement and interaction with the environment, letting tissue reorganization take place without interference. As the alarm phase finishes and moves to the resistance phase, the adapted tissue regains new levels of strength.

Consider again our blues singers coming from a low altitude to perform in a very high one. Let's say that when they arrive in Quito, they avoid overly stressing their cardiovascular and respiratory system by getting plenty of nutrition and rest. They practice singing in gradually increasing increments, so the odds are greater that the tissues will begin to adjust to the new, higher altitude and make performing a concert much easier. On the other hand, if they arrive in the Andes smoking cigarettes, exhausted from jet lag, and with poor baseline nutrition, the odds that exhaustion or outright tissue damage will ensue immediately increase.

✓ ***It is critically important not to add extra stressors to the stressed tissue during the Resistance Stage.***

In a nutshell, when our stress response kicks in, we are using a gift that lets us model our tissues to fit a new environment, but we are in a vulnerable transition phase. If we put more stress on already stressed parts, there is inevitable hell to pay in the form of damaged tissue.[23]

The stress response is there when you need it, but it comes at great cost. Just like with your financial situation, you are borrowing against the future. If you cannot resolve the stressor and get back to stable parasympathetic functioning, your body enters the stage of *exhaustion.*

The Exhaustion Stage

This is the stage where your biological credit card gets maxed out. The nerve terminals and glands start to run out of the adrenaline and norepinephrine needed to keep the heart rate and blood pressure up. However, under intense stress, as these concentrations increase, tissue exhaustion occurs and they become pro-inflammatory.[24]

> ✓ ***If the stressor persists and is not resolved, tissue and cellular damage occur. The ability to adapt lessens.***

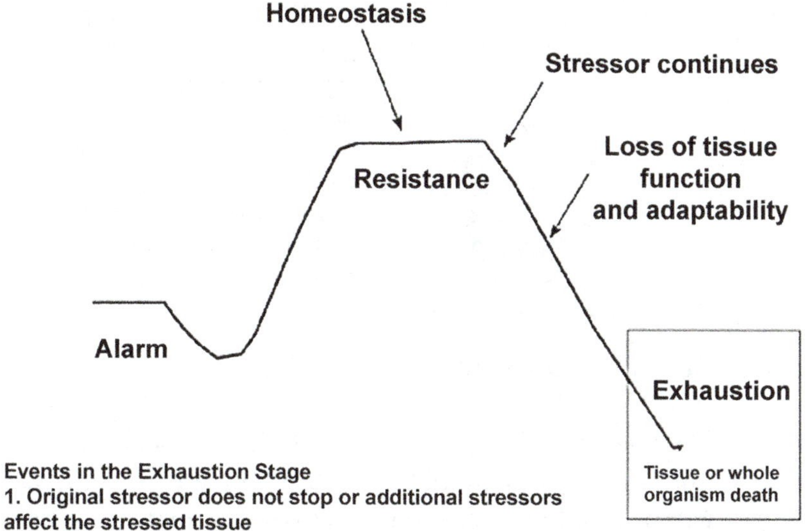

Events in the Exhaustion Stage

1. Original stressor does not stop or additional stressors affect the stressed tissue
2. "Adaptation energy" dwindles as HSP energy-intensive response mechanisms fatigue
3. Cortisol utilization has now depleted the tissue of building blocks and energy
4. Apoptosis or necrosis "cell death" starts due to inflammation resuming
5. In the early stages of exhaustion, there is loss of function and inflammation returns
6. The tissues become more defensive and rigid as they initially did in the alarm phase
7. Extracellular fluid collects, blood pressure drops, and waste and nutrient movement slow in the damaged area
8. If the stressor still persists, then tissue nonfunctionality or death ensues

Cortisol has gone beyond the stage of inhibiting inflammation and keeping tissues functioning. It is now directing your body to cannibalize muscle, skin, and other tissues to feed the heart and brain as a last-ditch effort to resolve the stressor. This exhaustion process can take place at the cellular, tissue, or whole organ level.

Back to our car crash: If you do not get help for that broken leg, when exhaustion arrives the damaged leg tissue will become necrotic. That is, it will actually start dying.

Another way to induce exhaustion is to put further stress on an already stressed body part too soon; for instance, running a marathon six weeks after a leg injury.

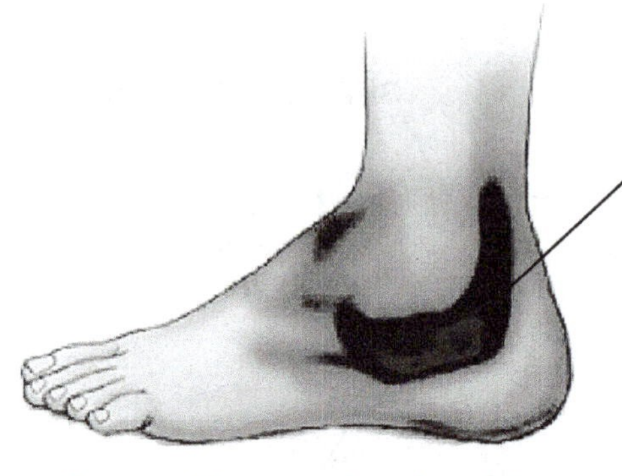

Foot with sprained ankle

This exhaustion phase is a central concept to this book, because it explains many of the illnesses that are caused by our modern way of life. We will be looking at this in greater depth shortly, so it may help to understand exhaustion a bit better. Stress is supposed to be an episodic phenomenon, not a chronic one. If it goes on too long the ever-so-helpful HSPs get fatigued, cortisol blocks energy from entering the cell, and this highly energy intensive process runs out of steam.

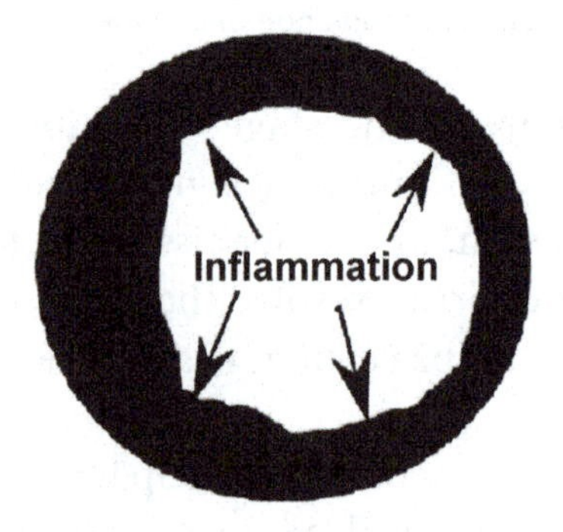

When inflammation spreads beyond stressed area to the rest of the body, shock occurs

At that point, the older inflammatory response, which is in part controlled by the parasympathetic system, kicks in and starts choking the tissue off from nutrients and waste removal in a big way. Shock (or inflammation), which was halted temporarily, restarts with full force. Hopefully it is just your leg or other local damage that will suffer, but if it is really bad, the trauma will go systemic and affect your entire body.

Tissue exhaustion leads to tissue dysfunction, or at the endstages, loss and death because the repair and stabilizing mechanisms become fatigued

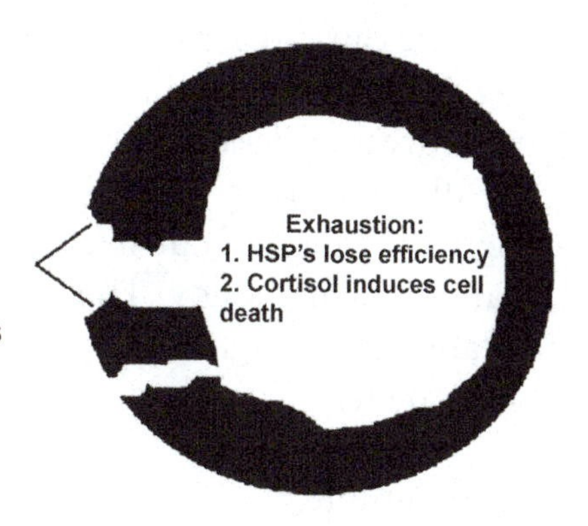

At this point, you have tissue dysfunction at the least, and maybe organismal death at worst. If it is just tissue, you will lose function. You might lose the leg, or have a limp and not be able to move around effectively. On a micro level, this is where cell aging occurs. On a chronic level, dependent on how the individual's genetic, personal habits, and environment are situated, this is the way every single disease (or pathology) is created. We will return to this crucial topic later.

> ✓ ***If the Exhaustion Stage persists, then irreversible tissue damage or death occurs. The stressed area resumes the inflammatory process and adaptation becomes difficult, if not impossible.***

The Key to Controlling Stress

What, then, is the key to controlling excessive stress? Never get to the point where you stay in the exhaustion stage for long periods of time. You must also avoid subjecting stressed tissue to another stressor before full recovery has occurred.

Another related point is to condition yourself in the alarm and resistance stages, when you are not facing an emergency to help build up your capacity for resistance. You would have more chance of avoiding tissue destruction in our car accident example if you were already in great athletic shape. Athletes use this capacity building concept all the time; it's called "training". We can apply capacity building to any part of our lives. Athletes train, the military does drills, and students study. The key is to extend yourself beyond comfortable limits, but not strain yourself to destruction from excessive fatigue.

We are not infinite. Selye (if he were still alive) would remind us that we have a limited amount of adaptation energy. Once it is used up, it's gone! Probably the biggest way to manage excessive stress is to prioritize our lives and evaluate which environmental stressors are worth embracing. Keep in mind that societies that have undergone the modernization process have, like clockwork, seen an explosion of stress related diseases – many of which were very uncommon 150 years ago.

What does this exactly mean? Well, I will posit that exposure to our Westernized lifestyle is directly resulting in many of the stress-related diseases with which we are so familiar. We will discuss this more in the next chapter.

Key Concepts:

- Identify the three phases of the General Adaptation Syndrome.
- Define the following terms:

Homeostasis
Shock
Apoptosis
Necrosis
Interstitial space

- What is the importance of each of the alarm, resistance, and exhaustion phases?
- What happens when fluid collects in spaces outside of the cells?
- What is meant by the concept of "temporary biological reset"?
- What is the difference between the sympathetic nervous response and the hypothalamic pituitary axis?
- What is the function of adrenaline (also called epinephrine) and cortisol during the resistance stage?
- Why is it critical to avoid additional stress on a stressed area during the resistance stage?
- How is the exhaustion phase similar to the alarm phase? How are they different?
- If you are an athlete who is in training, how much of your training should be spent in the exhaustion stage?
- At what stage of GAS does tissue death occur? At which stage does all disease originate?
- What is the primary key to controlling stress?

CHAPTER SIX

HOW THE BRAIN AND BODY CONTEND WITH STRESS

Chapter Objectives

- ✓ ***The prefrontal cortex is alerted during the Alarm Stage of the General Adaptation Syndrome. It can use imagination to avoid or quickly resolve problems, thus lessening the chances for cell and tissue damage in the Exhaustion Stage.***
- ✓ ***We are made to experience stress at irregular intervals, and not chronically.***
- ✓ ***Much of the processed food we eat creates intense chemical stress within our bodies.***
- ✓ ***We are made for cultivating social bonds and interacting directly with the natural environment.***
- ✓ ***The preventable stress we experience results in a huge economic drain to society.***

How Humans Adapt to Stress

We humans have a unique ability to alter our environment so that we can cope better with the stressors that come our way. Employing this ability throughout history has let us adapt to just about every harsh corner of the environment. If you think about it, it really is an amazing process.

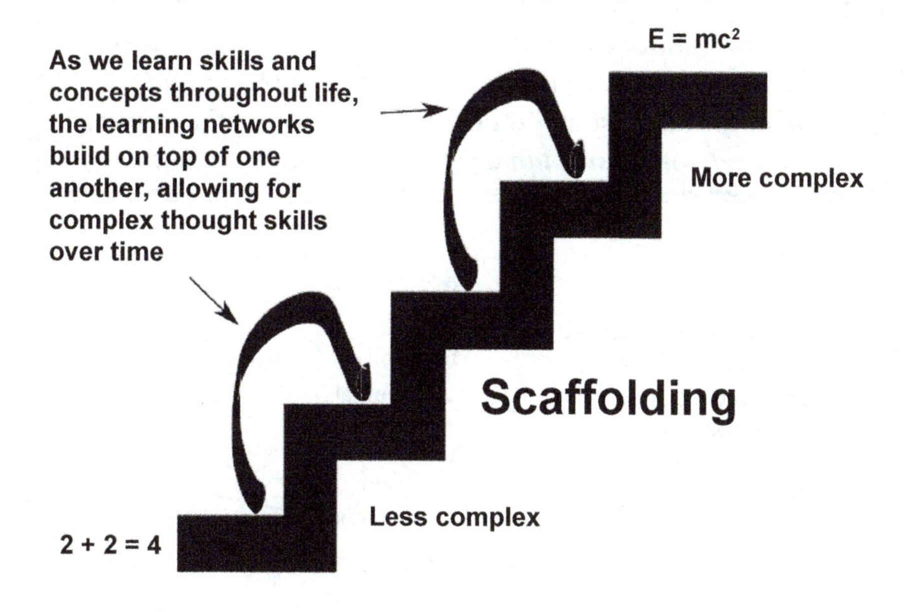

With our huge and freaky Frankenstein-like neocortex, we can form layers of associations in our mind running many magnitudes deep. Child psychologists remind us of the process of scaffolding, or the progressive build of concepts one on top of one another. For instance, it is one thing to understand that 2 + 2 = 4, but it is another entirely to have built concepts over many years that lead to differential calculus. The same goes for the basic speech of a two year old developing into eloquent poetry or the ability to speak multiple languages. We build our knowledge like climbing a ladder, adding more complexity with each step up.

This means that we can think deeply about things. Our prefrontal cortex, especially the part directly behind the eyes, helps us manipulate the environment by manipulating the future. We can spin numerous "as if" scenarios, or reenactments, at a very low metabolic cost. In this way, our imagination helps us evolve to adapt to the environment faster than evolutionary genetic mutations could ever manage.

✓ The prefrontal cortex is alerted during the Alarm Stage of the General Adaptation Syndrome. It can use imagination to avoid or quickly resolve problems, thus lessening the chances for cell and tissue damage in the Exhaustion Stage.

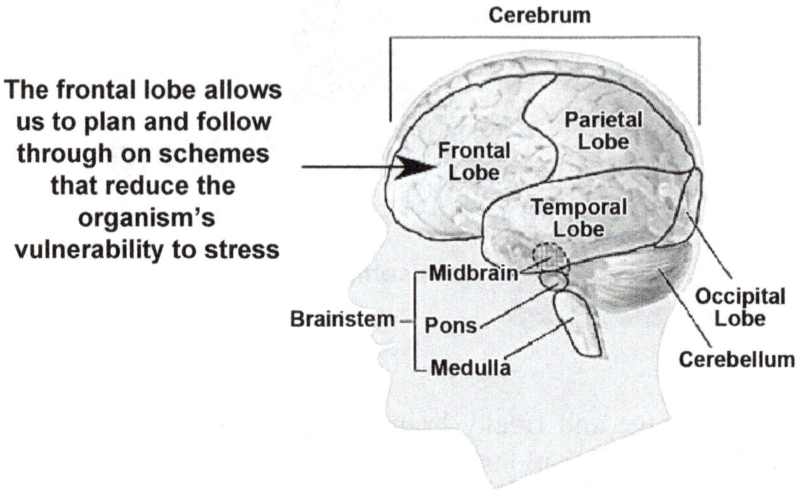

Cerebrum
The frontal lobe allows us to plan and follow through on schemes that reduce the organism's vulnerability to stress
Frontal Lobe
Parietal Lobe
Temporal Lobe
Midbrain
Brainstem
Pons
Medulla
Occipital Lobe
Cerebellum

Some of the many **prefrontal functions** include[25]:

- Language
- Novelty processing
- Creativity
- Temporal ordering of events
- Memory involved with verbal recall
- Metamemory (the ability to observe one's own thinking)
- Simulation (spinning "as-if" scenarios, brainstorming)
- Reality monitoring, which is proposed as a major mechanism in self-awareness
- Attention and orientation
- Actively maintaining information in working memory
- Changing behavior according to requirement of the environment

- Remembering clearly past events, and organizing current goals/future predictions
- Organizing and conceptualizing finances
- Integrating perception with action across time
- Attention to demanding cognitive tasks
- Modulating autonomic nervous activity
- Reward- and goal-related activity
- Retrieving information from long-term memory and metacognitive processes
- Regulating social and emotional behavior
- Emotional processing and behavioral self-regulation

With our imaginary scenarios in mind, we can plan and execute the work it takes to make our ideas become reality. It seems clear that our imagination was designed to reduce our exposure to stress by giving us temporary "work space" memory so that we have the bandwidth to make plans to alter the environment. In other words, our imagination is a type of a mental "sketch pad" that we can use without any external environmental stimulation.[26]

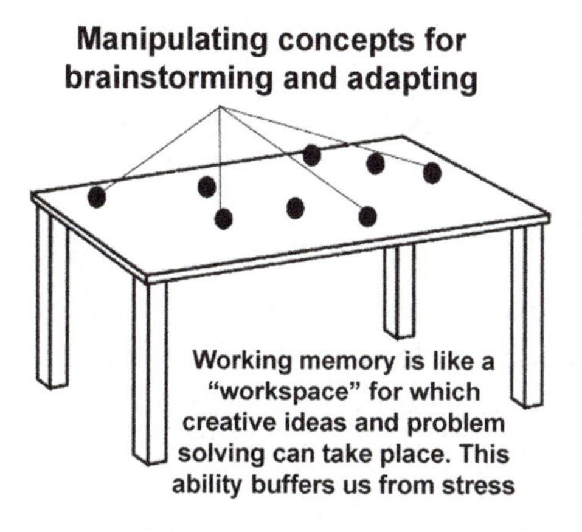

We briefly noted earlier in this book that by using the prefrontal cortex, we can build fires, sew clothes, and construct housing to avoid succumbing to cold stressors in Siberia. These basic functions sustained us for hundreds of thousands of years, but they are just the tip of a technological iceberg. It could be argued that our lives are now so governed by technology, piling new mental relationships onto older ones at lightning speed, that in the developed world, we are scarcely human anymore.

Nassim Taleb, perhaps the finest contemporary and most erudite of contemporary philosophers and the highly influential author of *Antifragile* and the *Black Swan*, gives us a stern warning about relying on systems that are too complicated. Especially ones that have been made too efficient, too large, too burdened, and without sufficient slack for error. Without some slack and simplicity, the complex system becomes too "fragile," thus increasing the odds of catastrophic damage and systemic failure.[27]

His line of reasoning equally applies to people. With our giant new brains (evolutionarily speaking), we have great potential to alleviate stressful stimuli, but also to create more stress. This arises from the sheer burden of carrying around such a huge, high maintenance, glitchy processing organ inside our skull. A Ferrari is a high performance, super specialized machine, but it is very high maintenance and vulnerable to the effects of mechanical stress. You cannot abuse such a machine without breakdown occurring. Perhaps it is wise to look at how we maintain ourselves in a similar light.

What's Going on in Our Heads?

We discussed earlier about how one branch of our nervous system processes the routine, cyclical elements of the world and another deals with the disruptive, novel elements. Our mental world is no different.

Many theorists and researchers have described stress in the context of danger and threat. While threats to existence are certainly stressful, they are only part of the whole story. It is the constant adjustment, the constant need to re-situate and reorient ourselves in response to stressors, that causes breakdown in our bodily system.

Major activation of our biological stress response is only designed to operate at *irregular* intervals. You see the lion chasing you, you flee or fight. Then if you are not eaten, you shake it off and get on with life. Hopefully, the times that you get chased by a lion are rare, as employing the stress response is a very expensive way of dealing with things. We are grateful it is there, but we should be very picky when spending our limited and valuable adaptation energy. Selye drove this point home over and over again in his writings – chronically engaging in situations that activate the stress response results in what he called the "disease of adaptation."

> ✓ ***We are made to experience stress at irregular intervals, and not chronically.***

But what about real, everyday life? We certainly are not chased by lions very often, so what is making us so vulnerable to stress related diseases?

The answer is that we lead very scattered, complicated lives in which we have chronic exposure to stress. Remember that complex systems are fragile. The more complicated our lives are, the greater the odds that we will need to activate our stress response over and over again in order to adapt and avoid going into shock.

This jibes with what our grandmothers always told us about slowing down. We don't have to be in dangerous situations all the time to develop cardiovascular disease. We just need to have constant low-level frustration, be messing with endless (and often irrelevant) details, worrying about things we cannot change, and engaging with constant self-stimulation through electronic devices or other tools of materialism. These activities and situations abuse our precious stress system's ability to generate adaptation energy.[28]

Our ancestral life was built around a fairly tranquil existence with episodic bursts of intense stress exposure. But the way nature put us together is very different from the way the modern world forces us to live.[29]

We evolved to have a slower, more uneventful way of living life, with the occasional need to lift heavy objects, to run, to keep up unrelenting vigilance during a hunt. We are *not* built for the daily grind of doing irritating mental work, sitting in traffic, brewing resentment or watching extreme violence via media or incessant gaming.

> ✓ ***We are made for cultivating social bonds and interacting directly with the natural environment.***

We are also not designed for coping with non-stop stimulation from the environment. Multitasking, having an excessive number of tasks to do in a limited amount of time, and being constantly hooked-up to our modern devices taxes our frontal lobe's ability to effectively imagine and execute useful scenarios when dealing with stressors.[30]

Stress from Processed Foods

Constant stimulation is not the only thing stressing our bodies. After the industrial and post-industrial ages, processed food became readily available. Not only that but, partly due to its extremely long shelf-life, it was often less expensive than non-processed meat, vegetables, and grain. As we have discussed previously, the more complex the system, the more vulnerable it becomes to unexpected shock.

> ✓ ***Much of the processed food we eat creates intense chemical stress within our bodies.***

Our ancestors consumed more of a raw, plant and/or meat based diet, although it is a matter of contention whether what foodies know as the "Paleo Diet" (few calories from carbohydrates) was really followed. It turns out that the gut bacteria we discussed earlier may have evolved to "fit" whatever diet was available.[31]

It's important to realize that much of processed food has chemical stressors. Let us explore this a bit further, starting with something as simple as French fries. A study looking at the composition of fries from a major fast food chain revealed nineteen ingredients.[32]Without going too specifically into the composition, such as the use of dimethylpolysiloxane, we can all readily agree that processed French fries are more than just potatoes and oil. Unknown to most people, there is a real metabolic cost to breaking down and digesting many of these industrially made chemicals. The liver and kidneys take the real hit, as trying to process or eliminate ingredients that our digestive tracts are not remotely adapted to is a huge chemical stressor to our systems. It is not just a stressor to our tissues, but likely a huge one for the functioning of our gut communities as well.

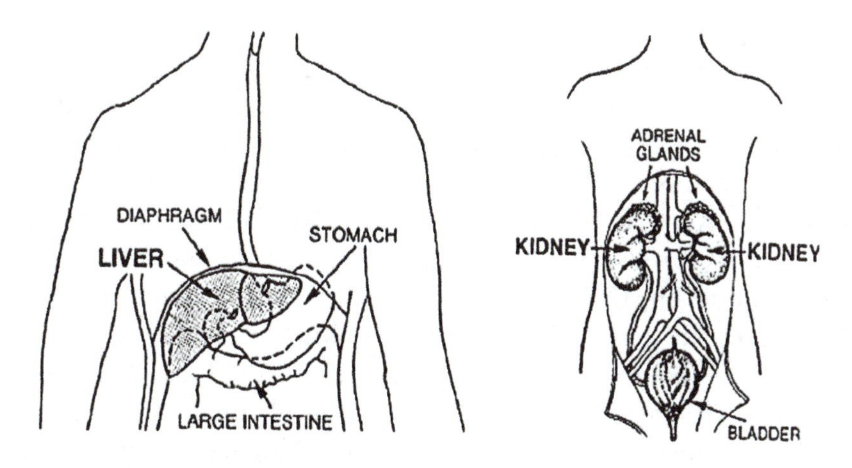

The liver and kidneys are vitally important for breaking down and removing cellular waste. The liver also participates with digestion and with immune functioning, while the kidney controls blood volume and blood pressure.

The constant stress of having to break down "weird" chemicals that were not present during our historical development as human contributes to the development of many common diseases.

Therefore, fast food, because of its non-native chemical composition, introduces significant stressors. These lead to significant disease formation in the developed world.

But wait, there's more! When we look at what we eat as an energy-in, waste-out model, it turns out that when comparing thinner hunter gatherers' energy consumption rate to Westernized US populations, it is roughly the same. More primitive cultures, with food consumption limited to simple and unprocessed foods, simply consume less calories.[33]

We need to reassess our concept of food in a new light. For argument's sake, let us suppose that most of what we eat is made of carbon atoms. If we consume so many carbon atoms that we exceed our digestive system's evolved carrying capacity, our bodies will interpret this as chemical stress and induce a robust stress response to avoid going into shock.

The same is said about the huge amounts of unnatural products we use for cleaning, repairing, or building our homes and offices. We use chemicals in gardening and agriculture and a variety of gels, lotions, and other potions for personal hygiene. The list of manufactured chemicals with which we surround ourselves goes on and on. If we as a society really want to understand the true financial, political, and human costs that we are paying by gratuitously activating our stress response, we need to look at all agents, whether psychological, chemical, or physical, that force our bodies to adjust excessively. The reasons why public policy should address this are many. One primary economic concern is that stress in developed societies results in a huge drag on gross domestic product.

> ✓ ***The preventable stress we experience results in a huge social and economic drain to society.***

In our next chapter, we look a little more closely at the price we are paying, both as a society, and as people, for all the stress that our modern world places on us.

Key Concepts:

- Define the following terms:

  Neocortex
  Scaffolding
  Working memory
  Metamemory

- What is the importance of the prefrontal cortex?
- What does the prefrontal cortex do to lessen our vulnerabilities to stress?
- Name five prefrontal cortex functions that alleviate the most stress in your personal life.
- Explain the significance of Nassim Taleb's work.
- Is stress only caused by a threat or intense fear?
- Why is exposure to stress only supposed to be episodic, and not chronic?
- How does processed food become a stressor to the body?

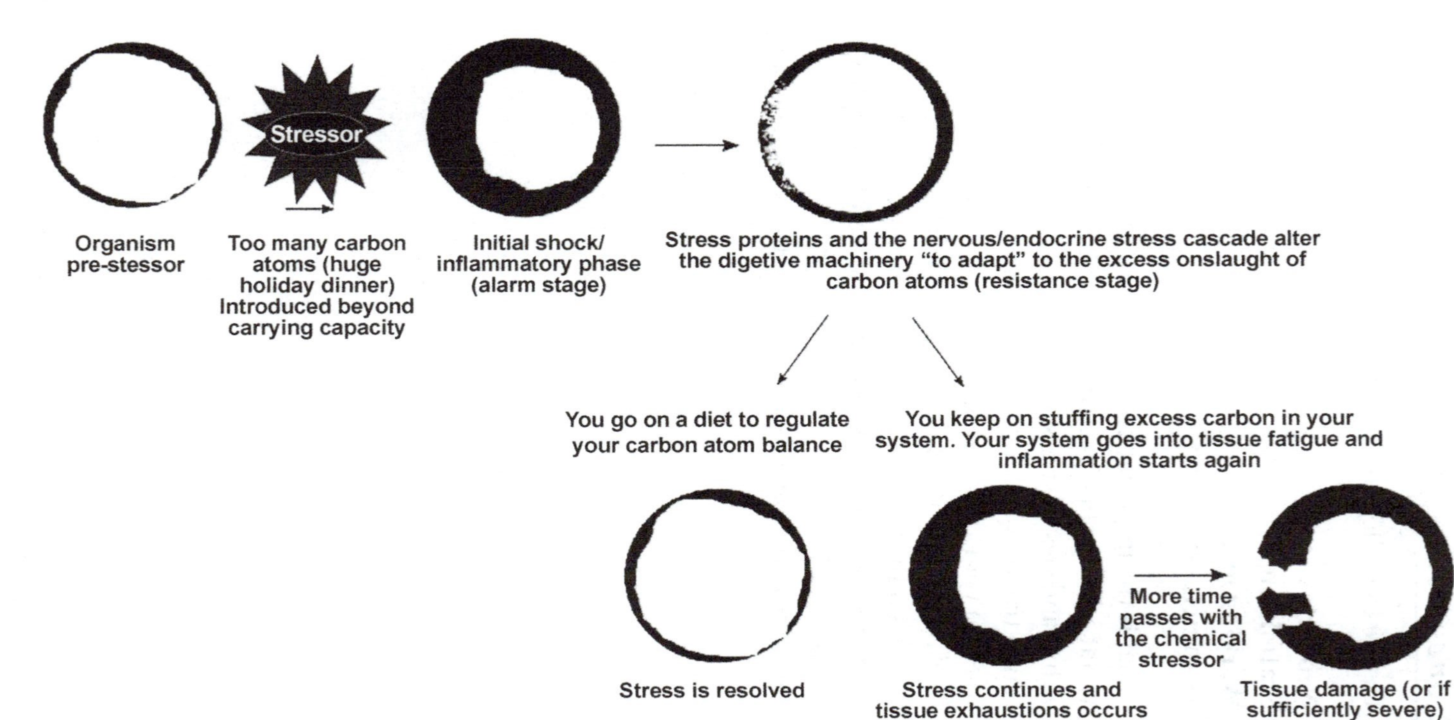
Stressor
Organism pre-stessor
Too many carbon atoms (huge holiday dinner) Introduced beyond carrying capacity
Initial shock/ inflammatory phase (alarm stage)
Stress proteins and the nervous/endocrine stress cascade alter the digetive machinery "to adapt" to the excess onslaught of carbon atoms (resistance stage)
You go on a diet to regulate your carbon atom balance
You keep on stuffing excess carbon in your system. Your system goes into tissue fatigue and inflammation starts again
Stress is resolved
Stress continues and tissue exhaustions occurs (exhaustion stage)
More time passes with the chemical stressor
Tissue damage (or if sufficiently severe) organism all death

CHAPTER SEVEN

THE STAGGERING FINANCIAL AND HUMAN COSTS OF STRESS

Chapter objective

✓ ***The preventable stress we experience results causes almost incalculable human and financial loss.***

Putting a Price on Stress

The cost to developed nations of unmitigated stress and trauma symptoms is staggering. Nevertheless, effective, rapid, and low-cost treatments are uncommon. The losses are in the multiple trillions of dollars worldwide each year. In 2011, the World Health Organization issued a report estimating the global cost of mental illness for 2010 at nearly $2.5 trillion, with a projected increase to over $6 trillion by 2030.[34]

Though exact and comprehensive costs are difficult to assess, a 2001 estimate put the cost of just workplace stress alone at $300 billion a year for the US.[35] In the EE-15 countries (Europe), this estimate ranges from 185 to 289 billion euros. It cannot be disputed therefore that unmitigated stress results in a massive drain on GDP for Western economies.[36]

The wealth of this evidence also suggests that the workplace could be an important contributor to the high health care spending and poor health outcomes in the U.S.[37] These costs do not reflect stress due to military combat or domestic abuse, nor the effects of stress from living in a hectic, over-scheduled, multitasking environment, or the costs of stress due to adverse childhood events (ACE), poverty, and inequality.

For US veterans seeking health care and disability assistance, the direct costs run over $6 billion a year. Of course, for the US, this is just a drop in the bucket. An estimated 8 million people, including veterans, have PTSD in any given year. The cost of treating anxiety disorders, including PTSD, in the US runs $42 billion per year.[38-39]

However, the true costs of dealing with unresolved trauma may be far higher, as cost from childhood abuse and neglect are estimated at over $585 billion a year in the US.[40] Another government study puts the price tag at $124 billion a year.[41]

Substance addiction, often the result of a person's attempt to self-soothe because of exposure to stressful events, whether chronic or episodic, is out of control in the US. It is estimated that the cost for substance abuse is about $700 billion per year.[42]

✓ ***The preventable stress we experience results causes almost incalculable human and financial loss.***

I proposes that in an age of economic deceleration, national policy makers could realize real savings and prevent public backlash by focusing on mitigating public stress rather than implementing financial austerity – which dramatically increases directs costs and the chances of public unrest.

Key Concepts:

- Roughly how much does exposure to stress cost worldwide per year?
- Are the costs of dealing with uncontrolled stress a threat to a nation's economic stability?
- How much is spent worldwide just on mental illness, according to the World Health Organization (WHO)?

CHAPTER EIGHT

THE IMPORTANCE OF THE HUMAN EMOTIONAL EXPERIENCE

Chapter objectives

- ✓ ***The most intense stressors we are experience are between each other as social beings.***
- ✓ ***Stress hormones can contribute to a person's social problems.***
- ✓ ***Our gut feelings arise from processing our sensory input over a long period of time.***
- ✓ ***Logical thinking requires your emotional input in order to create the most adaptive response.***
- ✓ ***Physical body sensations help to create emotions in our brains. Emotions partially reflect our sensory reality. If we refuse to recognize a person or group's feeling and emotions, we negate their physical reality. This is the cause of much social stress.***
- ✓ ***Childhood stress participates in the development of stress-related diseases later in life.***

The Complexity of Human Emotions

Emotions are weird animals. As rational as we would like to think ourselves, we are anything but. Western societies have a tricky relationship with the topic of emotions, because in order to appear professional and civilized, we tend not to encourage their expression. Unsuccessful parenting, failed relationships, and social unrest all share the fact that the emotional expression of the aggrieved party is not being heard. Avoiding this, by cultivating a developed sense of empathy, could be one of the most potent mitigates of the human stress experience.

However, emotional expression and regulation are not just neat neural replications of the environment. Poorly regulated emotional states are a major source of stress for humans. In fact, *relational conflict, which can result from unregulated emotions, is considered by many to be the most intense forms of stress.*[43]

This is worth emphasizing. The most potent form of human stress known is not running from a bear, sitting in traffic, or a myriad of other common examples. It is **relational stress**. This is not just the musing of feel-good pop psychology. Research in both humans and mice shows that empathy, which is possibly the most important single factor in cultivating successful societies, is blunted markedly by the effect of stress hormones.[44]

> ✓ ***The most intense stressors we are experience are between each other as social beings.***

Robert Sapolsky, perhaps one of the most influential thought leaders in stress research, says that the flooding of stress hormones in our brain can contribute to the genesis of anti-social behavior, such as xenophobia, a propensity to interpret unclear social signals as hostile and a tendency to lash out at those around us. Studies have shown the relevance of this to the economy, by demonstrating that stress makes participants in experimental economics games less generous and less cooperative.[45]

Not only can stress affect social cohesion in a society, but by blunting trust, it can profoundly affect economic output. Optimal emotional expression and regulation have profound effects on a society's emotional, physical, and financial health.

We have minimized the importance of our emotions by becoming cogs in a highly automated social machine. This is a grave mistake that goes against the very biological core of our being, as the reality of any situation involving human interaction ultimately depends on "how we feel about it."

✓ ***Stress hormones can contribute to a person's social problems.***

Ways to Reduce Relational and Emotional Stress

Any sort of social relationship between people – be it familial, platonic, romantic, business, or political – is much more likely to be successful if there is an empathetic response. When the feelings of another are disregarded, trust is lost, and a hostile response ensues. To engender success in relationships, business, or policy making requires an effort to understand and respond to the feelings of the other party. The alternative is tyrannical rule, which will always generate resistance. When there is resistance, the stress response gets activated, and when this happens, social dysfunction occurs and the costs of lost production and wasted resources skyrocket.

Evolution through Empathy

To see why empathy is one of the wisest and most evolutionarily important mechanisms for preventing relational upheaval, we need to turn to the work of two pioneers in emotional neural science, Antonio Damasio and Joesph LeDoux. They show that emotions (or feelings, depending on the level of consciousness) are the higher brain representations of body states. Let us dissect this.[46-50]

Many years ago, Antonio Damasio demonstrated that our emotions are actually involved in the rational decision making process. It turns out that the limbic system not only processes bodily states into feelings, but also records and remembers these body states during emotional events. The "gut feeling" you have in some situations is simply your limbic system giving the processed feeling felt from the last time you were in a similar circumstance.[51]

It is easy to understand that we have physical states like blood pressure, blood sugar level, or pH. We also talk about our temperature, water regulation, and muscular tension states (including the involuntary muscles of the gut and circulatory systems). We produce a myriad of physical states that change from moment to moment.

Information about many of these states is transmitted to the brain by their sensory input mechanisms, as discussed earlier, as well as by the blood. The information they send is detected by a tiny area of the deep and ancient brain structure clusters called the hypothalamus, specifically a part called the paraventricular nucleus. Here we have all sorts of barometers, thermometers, and other types of detectors picking up basic information about our body.

> ✓ ***Our gut feelings arise from processing our sensory input over a long period of time.***

In other words, the paraventricular nucleus is like a "real time" ticker tape recording our physical reality at any moment. The information it receives is processed by components of the higher brain called the *limbic system*. The conscious experience of a complex emotion, which develops from the raw data that *feelings* provide, is generated in other higher brain centers such as the frontal lobe.

Metaphor for the Somatic Marker Hypothesis

For instance, if you were once harmed by shady characters hanging out on street corners near Foch Plaza in Quito, then it is easy to see the utility of having a "reminder" when you see other shady characters in other cities. This mechanism is not 100% accurate, as racist views or other unfair scapegoating mechanisms and irrational phobias can occur. However, for getting an efficient gut feeling sample, which is often highly accurate, it is critical for making wise decisions in life.

Damasio noted that when patients suffer stroke or some other brain injury that results in damage to the underside of the frontal lobe - called the ventral medial and orbital frontal areas - conscious gut feelings disappear, and effective decision making and judgment go with them.

> ✓ ***Logical thinking requires your emotional input in order to create the most adaptive response.***

So there really is no such thing as an effective Spock from *Star Trek*. Without basic processing and retrieval of emotions, these patients are unable to identify gut "danger" signals, and they tend ultimately to gravitate towards shady characters and situations. In short, they are unable to learn from "emotional" mistakes. Emotions are therefore critically important for rational thought.

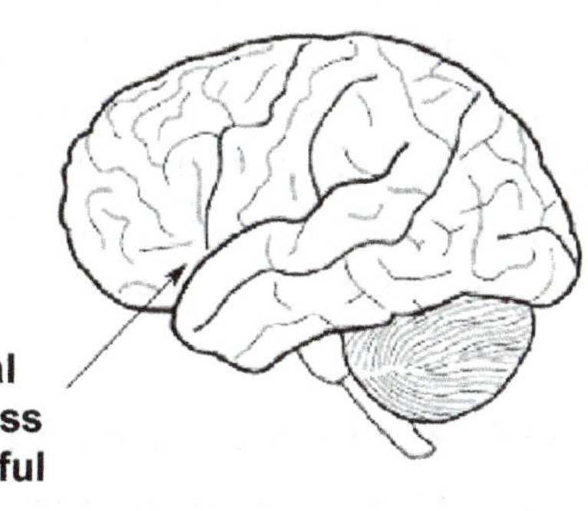

Ventral and orbital frontal areas process consciously stressful stimuli

It is important to note a profound consequence at this point. If our emotional experiences arise from physical and chemical data reporting on the body's operations, an emotional experience is in fact based on a person's physical reality - that is, on physical feedback from the body. Our emotions are therefore more than just 'in our head.' They are an expression of our physical reality.

The reason stress has such a profoundly damaging effect on social and emotional reasoning, is because invalidating or attempting to repress someone's feelings is actually negating their physical reality. This challenge naturally initiates a strong stress response in the person whose physical existence is being negated.

The primary cause of human socisl and stress related problems

According to Damasio and many others, our emotional feeling experience is a processed reflection of our reality. → **Huge amounts of relational stress*** are created when a person, group, or society negates emotional experiences of other persons, groups, or societies.** → **This leads to the development of stress related morbidity and mortality. It also greatly strains family, societal and political structures.**

*****relational stress is the most potent and damaging of human stressors**

When we negate and disregard a person's perception of their reality, we are in essence disregarding the whole human being, as we have just seen on actual measurement of physical states. Severe, protracted social stressors in history have arisen from the fact that people tend to disregard the welfare and feelings of other individuals, clans, and societies. Coercing or disregarding others by employing unyielding displays of power goes against the way in which humans naturally form bonds and cultivate networks. The short term gains it brings in terms of dominance and control come at profound social, political, and financial costs.

> ✓ ***Physical body sensations help to create emotions in our brains. Emotions partially reflect our sensory reality. If we refuse to recognize a person or group's feeling and emotions, we negate their physical reality. This is the cause of much social stress.***

Since we depend on social ties for survival from birth, being disregarded can result in a feeling of annihilation. This is often the root of debilitating psycho-social stressors. I propose that cultivating basic empathetic skills not only decreases the stress response in others around us, but also in ourselves, as the chances increase that those around us will stress us out less.

What Causes the Cost of Emotional Stress?

What causes emotions to become uncontrollable, and why does this experience result in so much cost in terms in financial loss and quality of life? To understand this, we need to look at the way we develop as babies, because our ability to regulate our emotions is closely related to the quality of the care we received. Emotional development comes before other advanced brain processes, like speech. To learn to regulate our emotions in a healthy way, we must develop in a healthy emotional environment. For this to happen, we need to form a bond with a caregiver who makes us feel safe and secure.

John Bowlby defined infant attachment as an emotional bond that ties the infant to one or a few figures across time and distance.[52-53] As infants, we are totally helpless and dependent on our caregivers. Although she may not understand the complexities of adult speech and emotional drama, a baby does instinctively know the difference between "safety" and "non-safety". If a parent is immature, stressed, or has little support, then tensions can be forced on the infant in the form of parental anxiousness, rejection, and other behavioral and physiological expressions of emotional instability.

Since the infant is totally helpless, and its innate sense of life or death rests in the hands of its caretakers, a huge amount of stress can develop as a result of the infant's attempts to negotiate its caretakers care and attention. In essence, because the caretaker is so unstable, the infant's newly forming emotional responses are formed based on fear and survival, not security and safety.

Constructing one's emotional development in life based on fear rather than safety has profound impacts on the adult's ability to emotionally regulate and to negotiate relationships. In fact it is hypothesized that healthy, functional, and long lasting romantic adult relationships are also based on mutual feelings of safety.[54]

Oxytocin is extremely important for bonding, especially between kin and social groups.

Oxytocin is a potent inhibitor of the sympathetic stress response. As a result, oxytocin is critical for modulating stress

The deep-seated, pre-verbal feelings of safety are thought to be mediated by oxytocin, a hormone released from the posterior hypothalamus, and implicated in the bonding process between caretaker and infant. More will be discussed in later volumes with the effects of childhood maltreatment, early stressors and the development of stress related diseases later in life and of shortened life spans. This factor determines whether a person can modulate stress successfully or poorly in infancy. Once established, to change the emotional habits that were established in pre-verbal, preconscious stages in later life as an adult, is very challenging but certainly not impossible.

> ✓ ***Childhood stress participates in the development of stress-related diseases later in life.***

Key Concepts:

- Define the following terms:

  Xenophobia
  Empathy
  Paraventricular nucleus

- Which type of stressor is the most potent and damaging for humans?
- Explain the significance of Robert Sapolsky's, Antonio Damasio's, Joseph LeDoux's, and John Bowlby's contributions for the study of emotions and stress.
- What is the significance of the Somatic Marker Hypothesis?
- How does stress affect human social functioning?
- How can a person reduce relational conflict within their social group and with their society?
- What are the functions of the ventral, medial, and orbital parts of prefrontal cortex?
- Can we make rational decisions without the input of our emotions? Could the character, Spock, from the TV series *Star Trek* really exist and still function in real life?
- How does safe environment affect an infant's emotional development?
- What is the importance of oxytocin for the regulation of the stress response and social bonding?

CHAPTER NINE

STRESS AND AFFLUENCE IN THE MODERN WORLD

Chapter Objectives

- ✓ *Emotional stress contribute greatly to the six leading causes of death.*
- ✓ *The current costs of stress are unsustainable for our society.*
- ✓ *Stress can negatively affect working memory performance, and conversely, poor working memory result in poor stress management, resulting in more stress.*
- ✓ *Previous exposure to stress can affect a person's future social standing.*
- ✓ *Inequality results in a vulnerability to the environment. This loss of control over one's environment results in highly increased disease and early death rates.*
- ✓ *Having a strong social support network is very important for lessening the impact of stress due to various causes.*
- ✓ *Social capital helps create livable, lower-stress societies.*
- ✓ *Biomes are functional extensions of our bodies. They are essential for our survival.*
- ✓ *Many of the products we use and the food we eat put heavy chemical stress on our bodies.*
- ✓ *40% of all deaths are thought to be caused by the chemical stressed caused by pollution.*

Development of Stress Related Disease Since the 1900s

To be human is to endure the stress response, as it shapes our bodies, as well as our experience in the world. The ancient Buddhist texts even had a term for stress: *dukka*. Dukka refers to the type of downstream problems that arise when initial conditions just do not "fit". A common analogy is when an axle does not fit snugly into a wheel, which makes all resulting motion jarring. Stress often begets more stress.

For those of us that live in the "modern" world, the rate of stress related diseases has exploded since the 1900s. Comparing data from 1900 to the year 2010 reveals that the two most common causes of death, heart disease and cancer have increased astonishingly, by 147% and 289%, respectively.[55] Furthermore, the occurrence of type II diabetes, which is essentially a disease resulting from the chronic chemical stress of our bodies, has increased by an astonishing 770% from 1958 to 2013.[56] The stressors in this instance is simply too many calories bombarding the digestive and endocrine systems.

Death by Stress

The disease and death rates due to stress-related illness is very disturbing. Emotional stress contributes greatly to the six leading causes of death in the US. These are cancer, coronary heart disease, accidental injuries, respiratory disorders, cirrhosis of the liver and suicide, respectively.[57]

> ✓ ***Emotional stress contribute greatly to the six leading causes of death.***

The developed world, especially the United States, consumes alarming amounts of prescription drugs, with almost 50% of the population taking one or more drugs within the previous 30 days.[58]

The rate of antidepressant and anti-diabetic pharmaceutical use in the EU has nearly doubled between the years of 2000 to 2010.[59] Furthermore, one in five men and one in four women have difficulty (associated with disability) in successfully completing basic actions associated with movement, sensory, emotional, or cognitive control.

There are many other statistics in the same vein, but this is enough to demonstrate that exposure to stress, and the problems that result from it, are a major financial and societal burden in developed societies. We need to examine whether the way we are living in more affluent nations is sustainable on a social, financial, and individual psychological level. In discussing other problems facing our society, we are ignoring the elephant in the room – the fact that people are facing physical and mental exhaustion due to the constant exposure to stress from living in the modern world.

The Problem with Affluence

It appears that simply being wealthy is not an automatic ticket to good mental health, especially among youth. Instead, wealth provides the means to indulge in a higher rate of self-soothing escapism to avoid relentless social and familial pressures.

Affluent teens have significantly higher rates of all major types of substance abuse compared to their lower income counterparts, including use of cigarettes, alcohol, marijuana, and hard drugs. The rates of clinical depression and anxiety show similar trajectories.[61]

We falsely assume that affluence and development equate to happiness, lower stress, lower disease and lower addictions rates. This is just not the case. Living in a country that is suffering grinding poverty is obviously very damaging. We must also closely look at the damage to humans that happens in developed societies.

It seems that development only goes so far to improve people's lives. For the downside, it can run up a huge bill for society for financing stress related disorders. As citizens and policy makers, it is imperative for us to search for the reasons why excessive affluence and technological advancement produce such damaging and catastrophically expensive outcomes.

These outcomes, if not urgently addressed, threaten to destabilize our entire economic and political system. The direct and indirect costs of avoidable stress top trillions of dollars every year. We cannot run an efficient society or economy, one that is capable of solving the problems of the future, with such a large percentage of the population incapacitated due to the effects of preventable stress. This amounts to hundreds of millions of people across the Western world losing the ability to be productive members of society.

> ✓ ***The current costs of stress are unsustainable for our society.***

If "development" leaves our entire Western civilization in danger of economic decline and social unrest, then we must identify the reasons behind this:

- Complexity
- Inequality
- Loss of social capital
- Destruction of personal microbiomes
- Chemical stressors

We will explore each of these in greater detail below.

Complexity

We have seen that the prefrontal cortex is extremely important in mitigating stress by allowing *planning* to take place. The imagination that comes from planning – the spinning of multiple "as if" scenarios for possible outcomes to a problem – made the advancement of civilization possible. The elaborate forms of government, economics and commerce, social organization and technology that we know today are built from the collective planning of millions of people. That is, millions of frontal lobes.

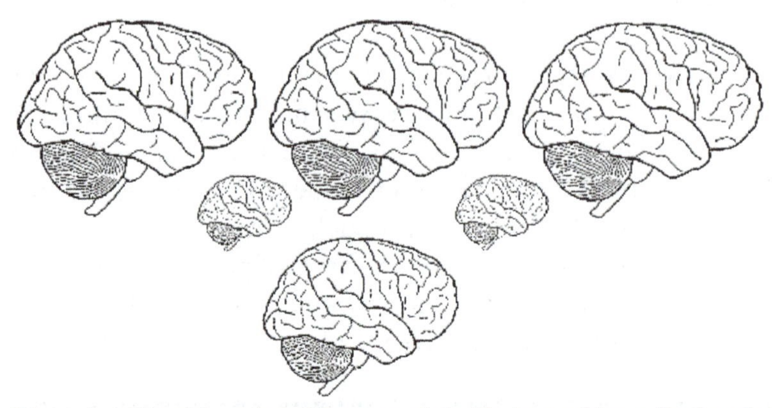

The adaptive advantage for a society that combines its frontal lobe capabilities are enormous. This greately reduces stress in a social system.

The major problem with this magnificent ability is the biological constraints of our planning muscles. Planning requires a lot of *working memory,* which is very similar to RAM in your laptop. Without working memory, we stop having complex human functioning with the outside and inside environment.

What is **working memory**? Working memory is like having access to a bunch of temporary Post-It notes in your mind. It lets you retain temporary details, like what you are wearing today, what you wore yesterday, the phone number to the pharmacy, or the list of tasks you need to complete today. You can manipulate its details fluidly and temporarily in order to complete tasks, but the slate gets wiped clean when the information is no longer needed.

As we discussed earlier, when we become stressed cortisol is released along with a cascade of other hormones, neurotransmitters, immune factors, and cellular proteins. Cortisol tends to be released more slowly that the other factors, and its purpose is to help tissues withstand the resistance stage of the General Adaptation Syndrome.

But the prefrontal cortex is packed with cortisol receptors, and the persistent elevation of cortisol over time has a strongly negative effect on its function. Many studies have demonstrated the impact it has in reducing our working memory.[62-64] This gets even more interesting if we turn it the other way around. Some people have a lot more working memory than others due to inherent differences, and it is thought that people with less working memory capacity are much more susceptible to stress than with those lucky enough to have a lot.

✓ ***Stress can negatively affect working memory performance, and conversely, poor working memory result in poor stress management, resulting in more stress.***

This suggests that ample working memory is a potent protector against mental stressors. Therefore, the prefrontal cortex is not only sensitive to stress, but when working properly, it actually protects a person from the effects of stress.[65]

These findings should not be taken lightly. The frontal lobes are what make us distinctly human. They are integral to our ability to adapt to a changing environment. If the demand for adaptation is too high relative to an individual's working memory capacity, then that person's negative reaction to stress skyrockets. What is this telling us?

How stress affects working memory

Working memory-pre stress

Thoughts, plans, and tasks

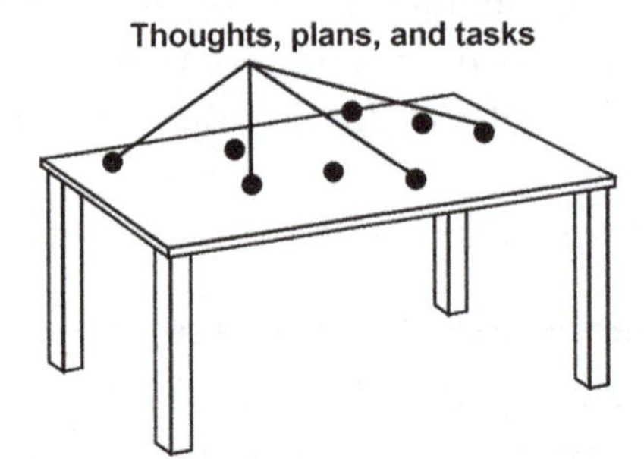

Working memory during the alarm and resistence phases of the adaptive response.

More thoughts, plans, and tasks

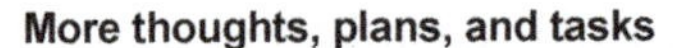

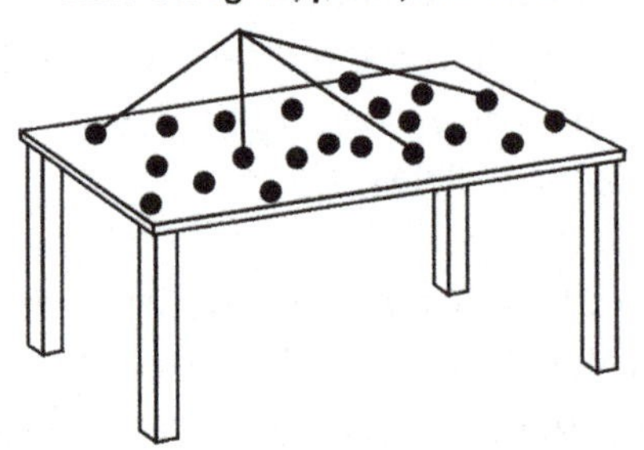

As working memory becomes strained. it enters into exhaustion stage and begins to become non functional. Cortisol is a big mediator of this process.

More thoughts, plans, and tasks to contend with at one time with shrinking working memory LESS WORKSPACE = loss of function

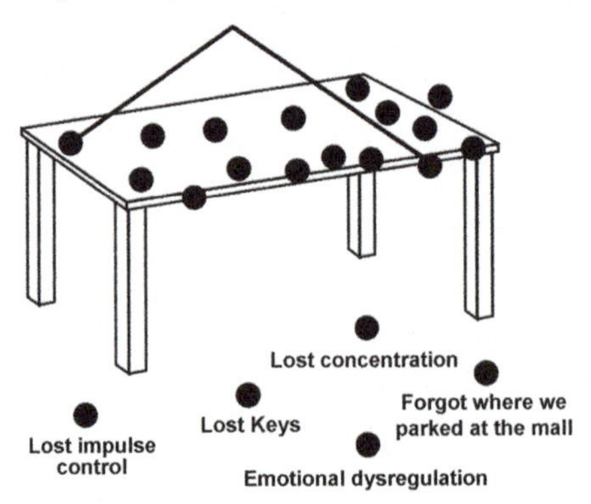

The sheer, chronic, everyday adjustments needed to cope with the thousands of decisions we must make that we did not have to make two hundred years ago, combined with the relentless stimulation from our electronic devices, do not go into a psychological void. They are either processed (at a cost), or valuable defensive processes (both psychological and physical) are expended to repel this input.

This is not an argument for throwing away all modern conveniences and finding a cave to live in. However, we must keep in mind that living in affluent, consumption-oriented societies puts more performance pressure on people's prefrontal cortices than is experienced by people in more traditional societies.

Inequality

Inequality of socioeconomic status has a clear effect not only on stress response in humans, but also on mortality. This is a bit of an odd duck, because although intuitively it is easy to understand a correlation between true material deprivation and early death, there is also a very strong connection between low socioeconomic status and early death within wealthy countries that have significant inequality. Despite the poverty being only relative, the health damage is thought to be caused by loss of social cohesion due to low positioning in the social hierarchy.[66]

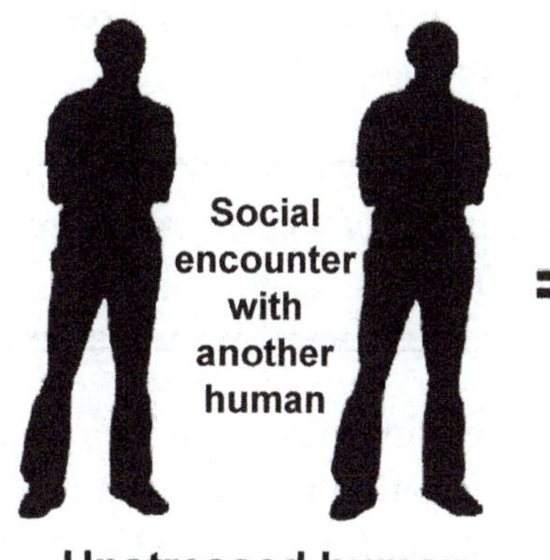

=

Possible rise in social hierarchy = less exposure and vulnerability to stress. Dominant members in a stable hierarchy tend to lower cortisol, as well as less stress related morbidity and mortality.

Carman Sandi's brilliant research with both humans and rats reveals something interesting about the relationship between stress and social hierarchy. With both humans and rodents, she found that prior stress exposure affects social positioning after encounters with members of the same species. Those who had earlier stressors before the social encounters tend to fall to submissive positions in the social rank. Those that had no previous stress tend to rise to dominance.[67]

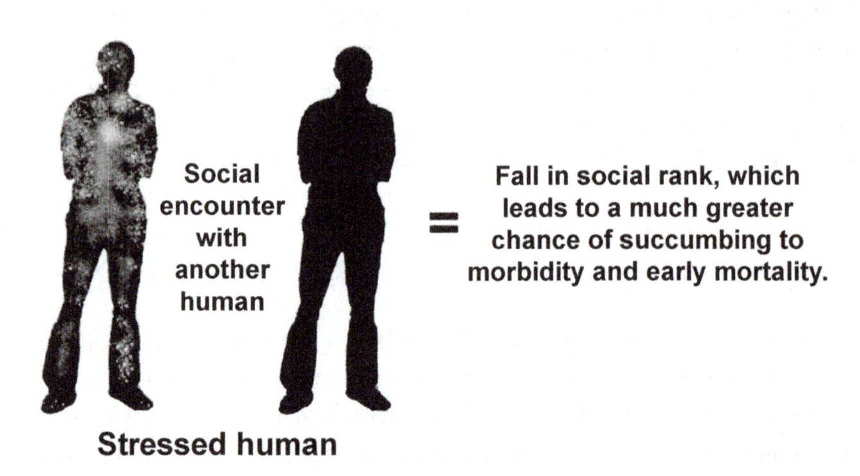

These findings are earth-shaking. What they tell us is that members of societies where there is inequality have a high likelihood of falling through the ranks if their social resources are exhausted due to previous stress. What this also suggests is that inequality may not create high stress and disease rates in a society per se. Rather, a stressful, possibly dysfunctional society that causes stress to its members will result in high levels of inequality.[68]

> ✓ ***Previous exposure to stress can affect a person's future social standing.***

Let us look at this more closely. In wealthy societies where there is large inequality, but true material deprivation is somewhat rare, yet we have this phenomenon of an increase in death rates and other stress related diseases. If true material deprivation is not the primary cause of the stress in this situation, then what is?

We must remember the definition of stress, which is the rate of adjustment it takes in order to adapt in a given environment. We must also remember that a healthy stress response makes our body shape and structure more plastic so that we can change ourselves to fit the environment, as altering the environment is not always possible.

When you are a low man on the totem pole, you are more at the whim of your environment. You basically have to be excessively plastic, ready to adapt to a myriad of environmental demands. A person with means and power has more opportunity to change her environment, rather than having to become more plastic to adjust to its demands.

Tiny elite—uses financial and social capital to bypass exposure to stress.
Greater control over environment.

Huge underclass—may or may not have access to social capital. Cannot use financial capital to "buy" oneself out of stress exposure. Must remain in the "plastic state" for excessive amounts of time. *Limited control over environment.*

We must remember that having flexibility is part of being human. Our systems are designed to cope with changes and stressful encounters, but only sporadically. When the demands of the environment are chronic and grinding, our mechanisms for adaptation become overtaxed and stop functioning efficiently, thus putting us at risk of multi-organ system failures. With too much plasticity, we lose structural integrity, and thus lose functioning.

> ✓ ***Inequality results in a vulnerability to the environment. This loss of control over one's environment results in highly increased disease and early death rates.***

Having little control over the environment presents tragic elements in human history. Whether the cause of this lack of control is war, epidemics, or simply brutal economic and political regimes, we can reliably see spikes in disease and death rates. When hierarchies become unstable due to excessive inequality, even alpha members suffer stress related ailments, as the advantages of being the leader of a sinking ship dwindle.[69] Thus gross inequality presents unstable situations, where everybody stands to lose.

Loss of Social Capital

Many studies indicate social support is an absolute must for maintaining physical and psychological health and overall resilience in the face of stress.[70] The importance of social capital for human functioning must never be underestimated.

Before the abstract construct of a financial economy was created in our society, there was the original economy that emerged from social relationships. The very complex financial economy that we live with everyday is something that has come to be seen as normal, natural, and even inevitable over centuries of human "progress." This is debatable, but one thing is for certain; our monetary system is not the original or even the most important economy. The original economy was (and is) one based, not on money, but on relationships.

We are immature when we are born and need many years of nurturing before we are ready to adapt to our environment, so we are evolutionarily wired to seek out social relationships.

As we mature and participate in our society, we learn ways to get our needs met. We find that social capital built through cooperation, bartering, and building goodwill by doing favors for each other greatly contribute to the odds of our survival.[71] From these dynamics emerge social capital. Wikipedia defines social capital as "the expected collective or economic benefits derived from the preferential treatment and cooperation between individuals and groups."[72]

> ✓ ***Having a strong social support network is very important for lessening the impact of stress due to various causes.***

Social capital therefore encompasses the goodies you get when you have a good support network. The definition of capital in general involves having some sort of capacity for storing assets or advantages for future use.

Financial capital is the common store of wealth in developed societies, and it does have its advantages. Money is an interesting animal in that it can be manipulated in many abstract ways, and can be exchanged for many different goods and services.

However, before the advent of financial systems, ancient societies had their own stores of wealth and processes of economy. This basic economy emerges when trust is cultivated between individuals. Bartering, trading favors and gifting can directly take the place of monetary exchange. As trust is cultivated between groups over many generations, the type of stable wealth that is generated for its members is substantial. Societies with stable mechanisms of social capital production have higher civic involvement, and with this involvement, they have more successful communities, as well as a better functioning government.

Indeed, the rates of participation in civic activities in the US has dropped precipitously over the last 50 years. Lack of social capital via our social supports is clearly associated with increased levels of stress[73.] Conversely, having ample amounts of social capital allow stressed people to bounce back to normal more rapidly.[74] Not only that, but in the US, being on the low-end of income inequality is strongly associated with loss of social trust and failure to identify with social groups.[75]

✓ *Social capital helps create livable, lower-stress societies.*

We have already seen that being part of a developed society may not automatically be beneficial for one's health.[76] Although there has been ample speculation as to why loss of social capital results in such startling increase of the stress related disease and death rates in a society with relatively less absolute material deprivation, concrete answers are still elusive.

Against this background, description of life in a so-called "less developed" economic structure, such as Ecuador, may offer a few answers. As hard as it is to imagine, Ecuador has had 20 constitutions, 10 forms of currency, and over 70 presidents since 1830, including a stint in the 1970s being run by the navy.[77-79]

Without being controversial, Ecuador's past can be described as unstable. As a response to incessant political and economic vacillation, it appears that Ecuadorians have maintained their strong sense of old world social cohesion and democracy, which has enabled them to resist many of the negative effects of modernization along with its ever-present companion, stress. Because efforts at modernization came very late in the game due to its political and economic instability, social capital remains high.

Even though Ecuador is officially classified as a middle class country by world standards, homeless and prison populations are just a fraction of its wealthier neighbors'. The people have been conditioned over many generations to gather in social and religious groups and to protest if they feel their rights are being violated. Civil participation, as well as greater generalized social activity, dominate society.

Thriving farmers' markets, that have survived for hundreds of years, operate outside of the formal distribution systems that supply much of the nation's food. The culture encourages small business ownership in a major way, though there are a few mega-corporations that run major media outlets and supply basic infrastructure. There is a large informal economy that has evolved to sustain society during currency crises and banking holidays. The people have prospered economically in recent years, however a healthy suspicion of government is almost ingrained into their genetic structure.

As noted, because of their long history of national instability, the Ecuadorians have a strong inclination to assemble in groups for civic and political organization. These networks form rich stores of social capital that sustain their culture outside of any artificially constructed political or economic system. Social capital functions as *safety insurance* in a profound way for people, lessening the need for stressful adjustment to the environment.

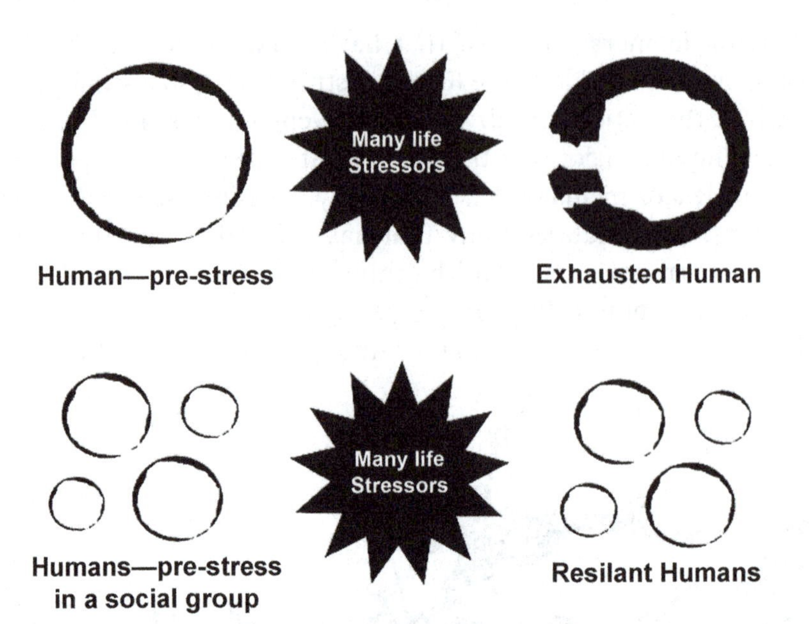

A social capital unit employs the inherent ability of its people to band together to influence the environment, whether it develops naturally from within the group, or is imposed artificially from without.

A person without access to this type of wealth has a limited ability to avoid stressful environments. She must spend large amounts of energy adapting to her in a vulnerable dog-eat-dog way, without the moral or practical support social capital brings. Just as people in unequal societies must be extremely plastic to survive, *living without social capital requires a highly plastic stance in order to contend with the repeated stressors the environment presents.* This forced plasticity puts a strain on the individual, thus greatly increasing the chance that stress related diseases and death rates occur.

Destruction of Human Microbiomes

We have discussed the intensely human issues of exhaustion due to working memory strain, social inequality, and the loss of social capital. Another cause of excessive stress and adjustment in developed societies is the loss of functional biomes.

Healthy Organism

We "pay" for the biome by becoming more plastic and not eliciting a defense response

A healthy organ such as the skin or the gut

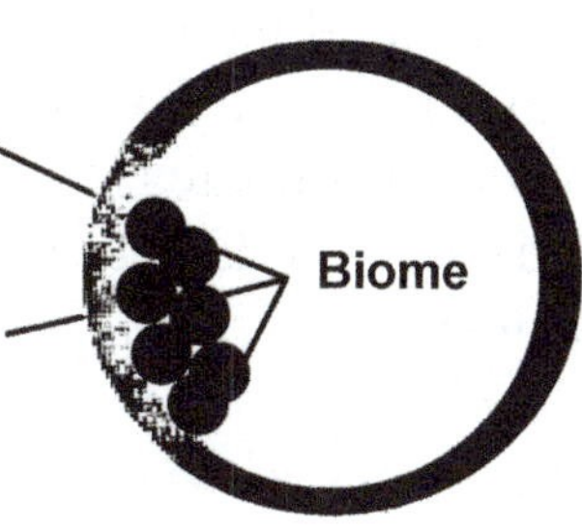

What we get in return for hosting good bugs

- Digestion of molecules
- Immune system development
- Defense against infections
- Synthesis of viatmins
- Fat storage
- Blood vessel growth regulation
- Behavior development
- Regeneration of the skin

As noted previously, hundreds of billions of microorganisms inhabit our skin, hair, gut, and orifices, and these creatures serve as functional extensions of ourselves.[80] The sum of organisms living on the surface and inside the body is termed **microbiota**. This environment is dominated principally by anaerobic bacteria, but also includes viruses, protozoa, archaea and fungi.[81]

Some have termed the collective unit of the host animal and its friendly organisms as a "superorganism". This is an individual's microbiome, and it is a dynamic entity that can be influenced by factors such as age, diet, hormonal cycles, travel, therapies, and illness.

> ✓ ***Biomes are functional extensions of our bodies. They are essential for our survival.***

The microbiome's importance to this book stems from the power it brings to human functions. These include polysaccharide digestion, immune system development, defense against infections, synthesis of vitamins, fat storage, blood vessel growth regulation, and behavior development.[82] It is generally accepted that our microbiota control the regeneration of the epithelium (the outer layer of the skin) and recruitment of various leukocytes (white blood cells) into the skin for greater host protection.[83]

In rodents, gut microbiota appear to influence the development of emotional behavior, stress and pain control systems, and brain neurotransmitter systems.[84] The microbiome's effect on human functioning must not be underestimated. Refer to the list below for diseases associated with depleted or deranged microbial flora[85-103] – a condition known as **dysbiosis**:

- Inflammatory bowel disease and Crohn's disease
- Nonalcoholic fatty liver disease, steatohepatitis, alcoholic liver disease, and cirrhosis
- Schizophrenia and Celiac disease
- Obesity
- Necrotizing enterocolitis (in pre-term infants)
- Allergic sensitization, allergic rhinitis, and peripheral blood eosinophilia (in school aged children)
- Dermatitis
- Multiple sclerosis
- Thyroiditis
- Atherosclerosis
- Inflammatory conditions in the vaginal tracts
- Chronic Obstructive Pulmonary Disease (COPD)
- Hypertension
- Old age related diseases
- Respiratory outcomes in Cystic Fibrosis
- Lupus
- Colorectal cancer

With research in the field of microbiome-disease connection exploding, promising results have indicated the possible use of probiotics, or introducing tissue-specific foreign microbe communities for the treatment of inflammatory diseases.

However, this discussion begs the question: *which aspects of the modern lifestyle result in depletion or outright destruction of the human biome in the first place?*

A major culprit, not surprisingly, is the widespread use of antibiotics.[104] A second likely factor is the typical Western diet. Switching from a low-fat, plant rich diet to a high-fat, high-sugar Western diet changed the structure of the microbiota within a single day in mouse models.[105]

A third disrupting factor to the human biome is the exposure to environmental chemicals and toxins. Through depleting certain key microbial communities, endocrine disruption occurs. For instance, in areas with arsenic contamination in the local water supply, there is a commensurate increase in Type II diabetes. Also included as disrupting suspects are Bisphenol A (BPA), alkylphenols nonylphenol and octylphenol, diethylstilbestrol (DES), and genistein.[106]

A fourth factor is maintaining overly sanitary conditions. David Strachan described the "hygiene hypothesis" phenomenon for which he posits that lower incidence of infection in early childhood could be an explanation for the rapid rise in autoimmune diseases such as asthma and hay fever in the 20th century.[107]

It is now also recognized that overly sanitized conditions contribute to a much broader range of chronic inflammatory diseases, which includes diseases such as type 1 diabetes, multiple sclerosis, some types of depression, and cancer.[108-111]

The important point to take away from this section is that the microbiome is a functional extension of ourselves, performing many functions necessary for life, and that there are numerous factors capable of disrupting its healthy function. This disruption makes us vulnerable to stressors which otherwise we might be able to resist.

Chemical Stressors

Our exposure to chemicals, and therefore our exposure to chemical stressors, has grown dramatically since the dawn of the industrial age. The increase in chemical stressors is tied directly to the loss of biome, which is itself a type of chemical stressor. The major sources of chemical stress are:

1. Substances we choose to consume, apply to our bodies or use in our dwellings; and
2. Substances we are involuntarily exposed to from the environment

Voluntary Consumption

Our bodies, and specifically our cellular stress mechanisms, evolved millions of years before the discovery of many of the chemical substances we now use. As a result, we are simply not adapted to deal with them.

The importance of this fact can be highlighted by considering that in just the last 50 years alone, 80,000 chemicals have been developed, used, distributed and discarded into the environment. And most of these have never been tested for potential toxic effects in humans or animals.[112]

However, perhaps the most potent chemical stressor is actually the food we eat, both in terms of quantity and quality. Our bodies are effectively food processing machines, and they have a limited capacity. If we consume more calories than we can process, our digestive and metabolic mechanisms become stressed. Stressing a system chronically, as we have already seen, results in disease.

> ✓ ***Many of the products we use and the food we eat put heavy chemical stress on our bodies.***

This is a curious predicament. From evolving to face threats from famine and hunger, technology has eradicated these threats over time and replaced them with a threat stemming from an overabundance of cheap, processed calories. These foods are high in refined carbohydrates, like white flour or sugar, which are not natural foods, and which themselves set off stress related cascades in the brain and body that can result in food addiction.[113]

Although it has been said before, it is worth emphasizing that the stress induced by the depletion of the gut biome and from the consumption of processed foods cannot be overstated. It is a major a source of disease formation.

Though a thorough discussion of this topic would take many volumes, a major toxic food stressor is none other than sugar. A fascinating study was conducted by replacing children's dietary sugar consumption with starches, so that the quantity of calories from sugar made up no more that 10%. After just nine days, the children's fasting blood sugar levels dropped by 53%, along with significant declines in insulin, triglyceride, LDL, and fatty liver levels. What is important to note is that the diets fed to the children were not necessarily healthy – they simply replaced the sugar with carbohydrates. Yet this one change made the children's metabolic profiles improve quickly and dramatically.[114-115]

Similar considerations apply to the drugs we take, and the chemicals we use for cosmetic or hygiene purposes. Their use may provide short-term benefits, but as with processed food, our metabolic machinery has not evolved to deal with these non-native substances. As many of them have been developed in just the last two or three generations, we simply don't yet know what the implications of long term exposure will be. However, we can assume that exposure to at least some of them will trigger the stress response. This risk makes their repeated use unwise.

Involuntary Exposure

Having looked at the substances we consume by choice, we will now look at those to which we are exposed from our environment. Hazardous chemicals escape to pollute the environment from numerous human activities, and many of these are capable of damaging our health. Major types of **pollution** that can be regarded as chemical stressors for humans include air pollution, soil contamination, and water pollution.[116] In fact, a Cornell study suggests that 40% of deaths worldwide are caused by these three.[117]

> ✓ ***40% of all deaths are thought to be caused by the chemical stressed caused by pollution.***

For instance, air pollution affects a number of systems and organs, with some of its harmful impacts including upper respiratory irritation, chronic respiratory and heart disease, lung cancer, acute respiratory infections in children and chronic bronchitis in adults, aggravating pre-existing heart and lung disease, and asthmatic attacks. Little wonder that air pollution alone is responsible for at least eight million deaths per year.[118-119]

Key Concepts:

- How have the death rates of cancer and heart disease changed since 1900?
- What are the five major causes of stress in the modernized world?
- Explain the significance of Carman Sandi's work describing the relationships between stress and social status.
- How has the incidence of diabetes changed since the 1950s?
- What type of stressor is a major cause of death?
- In the United States, what percentage of people have taken prescription drugs within the last 30 days?
- Which health risks are affluent adolescents most vulnerable to?
- Is it possible to finance a society in a sustainable manner if a large percentage of its people are incapacitated by stress related diseases?
- How does working memory help us deal with stress?

- How can stress and cortisol disrupt working memory? What are the consequences of this?
- How does inequality and loss of social capital make an individual more susceptible to stress?
- By way of contrast with the typical Western system, how does Ecuador's more ancient social and informal economic system protect its people from many types of stress?
- What are the two types of chemical stressors?
- What are seven ways a properly functioning biome can serve us? How is it that a biome can be a functional extension of our bodies?
- What are seven diseases that result from a poorly functioning or depleted biome?
- What are four reasons a biome can become depleted or malfunctioning?

CONCLUSION

I have given an overview of important factors contributing to lifestyle and stress-related disease and death in contemporary Western society. This is not intended to be a blanket condemnation of the Westernized lifestyle. Life in more primitive societies certainly exposes people more directly to the elements of nature, and introduces life-or-death vulnerabilities that we are not faced with in the West. There are good reasons why humans pursued the path of controlling our environment, and thus have developed non-indigenous technologies and lifestyles. Life in a primitive setting can be brutal and harsh.

Even Henry David Thoreau, writing in the 1800s, was careful not to totally condemn the positive aspects of modern technology. The key seems to be in maintaining a simplified ecology, that is, a human friendly environment demanding appropriate rates of adjustment. In some instances the fruits of modern society can fulfill this.

But if stress is simply the rate of adjustment needed to adapt to a particular environment, and the way our culture is structured is causing expensive and dangerous outcomes, then addressing the factors mentioned above should result in dramatically improved outcomes in terms of stress reduction and therefore our health, as well as the health of our political and economic systems.

Perhaps it is time to explore sustainable living as a matrix of both primitive and modern elements – to pick which elements of each serve us best, and to build a more satisfying, healthy existence.

Understanding the basic concepts of the human stress response and how overuse of its adaptive mechanisms is the cause of needless human suffering is the first step towards improving our lives. Along the way, we could reduce the devastating cultural, political, and financial losses to societies that have adopted modern ways of living.

Remember, you do have the power to change things.

REFERENCES AND RESOURCES

1. McEwen, Bruce S. "Allostasis and Allostatic Load: Implications for Neuropsychopharmacology." *Nature.com*. Nature Publishing Group, n.d. http://www.nature.com/npp/journal/v22/n2/full/1395453a.html

2. Csermely, Peter. "Plasticity - Rigidity Cycles: A General Adaptation Mechanism." *COMPLEX SYSTEM ADAPTATION Plasticity-rigidity Cycles*: n. pag. Cornell University Library. <http://arxiv.org/ftp/arxiv/papers/1511/1511.01239.pdf

3. Chan, Serena. "Complex Adaptive Systems." *Complexity, Institutions and Public Policy* n. pag. ESD.83 Research Seminar in Engineering Systems. <http://web.mit.edu/esd.83/www/notebook/Complex%20Adaptive%20Systems.pdf

4. Wenner, Melinda. "Humans Carry More Bacterial Cells than Human Ones." *Scientific American*. 30 Nov. 2007. http://www.scientificamerican.com/article/strange-but-true-humans-carry-more-bacterial-cells-than-human-ones/

5. Department of Experimental, Diagnostic and Specialty Medicine, University of Bologna. "Result Filters." *National Center for Biotechnology Information*. U.S. National Library of Medicine, n.d. http://www.ncbi.nlm.nih.gov/pubmed/23829164

6. "Skin Flora." *Wikipedia*. Wikimedia Foundation, 27 Feb. 2016. https://en.wikipedia.org/wiki/Skin_flora

7. "Infections Caused by Pseudomonas Aeruginosa." *Polyphor* -. 25 Mar. 2015. http://www.polyphor.com/products/pseudomonas-infections/infections-caused-by-pseudomonas-aeruginosa

8. Tauber, Alfred. "The Biological Notion of Self and Non-self." *Stanford University*. Stanford University, 21 May 2002. http://plato.stanford.edu/entries/biology-self/

9. Recordati, Giorgio. "A Thermodynamic Model of the Sympathetic and Parasympathetic Nervous Systems." *Autonomic Neuroscience: Basic & Clinical*. 31 Jan. 2003. <http://www.sciencedirect.com/science/article/pii/S1566070202002606

10. Martelli, D., S. T. Yao, M. J. McKinley, and R. M. McAllen. "Reflex Control of Inflammation by Sympathetic Nerves, Not the Vagus." *The Journal of Physiology*. BlackWell Publishing Ltd, 1 Apr. 2014. http://www.ncbi.nlm.nih.gov/pmc/articles/PMC3979618/

11. Ghacibeh, Georges A., Joel I. Shenker, Brian Shenal, Basim M. Uthman, and Kenneth M. Heilman. "Effect of Vagus Nerve Stimulation on Creativity and Cognitive Flexibility." *Epilepsy & Behavior*. June 2006. <http://www.epilepsybehavior.com/article/S1525-5050(06)00084-9/abstract

12. Recordati, Giorgio. "A Thermodynamic Model of the Sympathetic and Parasympathetic Nervous Systems." *Autonomic Neuroscience: Basic & Clinical*. 31 Jan. 2003. <http://www.sciencedirect.com/science/article/pii/S1566070202002606

13. Institute of Biochemistry, Hungarian Academy of Sciences Biological Research Centre. "Result Filters." *National Center for*

Biotechnology Information. U.S. National Library of Medicine, Dec. 2005. http://www.ncbi.nlm.nih.gov/pubmed/16302971

14. Fulda, Simone, Adrienne M. Gorman, Osamu Hori, and Afshin Samali. "Cellular Stress Responses: Cell Survival and Cell Death." *Cellular Stress Responses: Cell Survival and Cell Death*. 4 Aug. 2009. http://www.hindawi.com/journals/ijcb/2010/214074/

15. Lindquist, S., and E. A. Craig. "HEAT SHOCK PROTEINS EXPRESSION - Methods." (1988): n. pag. Massachusetts Institute of Technology. <http://lindquistlab.wi.mit.edu/PDFs/LindquistCraig1988ARG.pdf

16. Pockley, A. Graham. "Heat Shock Proteins, Inflammation, and Cardiovascular Disease." *American Heart Association*. <http://circ.ahajournals.org/content/105/8/1012.full

17. Balogh, Gabor, Iboyla Horvath, Eniko Nagy, Zsofia Hoyk, Sandor Benko, Olivier Bensaude, and Laszlo Vigh. "The Hyperfluidization of Mammalian Cell Membranes Acts as a Signal to Initiate the Heat Shock Protein Response - Balogh - 2005 - FEBS Journal - Wiley Online Library." *Febs Press*. 9 Nov. 2005. http://onlinelibrary.wiley.com/doi/10.1111/j.1742-4658.2005.04999.x/full

18. Schmitt, E., M. Gehrmann, M. Brunet, G. Multhoff, and C. Garrido. "Intracellular and Extracellular Functions of Heat Shock Proteins: Repercussions in Cancer Therapy." Journal of Leukocyte Biology. <http://www.jleukbio.org/content/81/1/15.full

19. Selye, Hans. "The Stress of Life." *WorldCat*.Web. 03 Mar. 2016.

20. Porges, Stephen W. "Orienting in a Defensive World: Mammalian Modifications of Our Evolutionary Heritage." n. pag. 1995. <http://condor.depaul.edu/dallbrit/extra/psy588/Orienting%20in%20a%20Defensive%20World.pdf

21. Hatoko, Mitsuo, Hideyuki Tada, Masamitsu Kuwahara, Tsutomu Muramatsu, and Toshihiko Shirai. "Epinephrine Induces 72-kD Heat Shock Protein (HSP72) in Cultured Human Fibroblasts." Jan. 1996. <https://www.researchgate.net/publication/40620453_EPINEPHRINE_INDUCES_72-kD_HEAT_SHOCK_PROTEIN_HSP72_IN_CULTURED_HUMAN_FIBROBLASTS

22. Murphy, SJ, D. Song, FA Welsch, DF Wilson, and A. Pastuszko. "The Effect of Hypoxia and Catecholamines on Regional Expression of Heat-shock Protein-72 MRNA in Neonatal Piglet Brain." *National Center for Biotechnology Information.* U.S. National Library of Medicine, 15 July 1996. http://www.ncbi.nlm.nih.gov/pubmed/8842392

23. DeMeester, Susan L., Timothy G. Buchman, and J. Perren Cobb. "The Heat Shock Paradox: Does NF-κB Determine Cell Fate?" *The Faseb Journal.* <http://www.fasebj.org/content/15/1/270.full

24. Pongratz, Georg, Straub, Rainer H. "The Sympathetic Nervous Response in Inflammation." Arthritis Res Ther. 2014;16(504).

25. Siddiqui, Shazia Veqar, Ushri Chatterjee, Devvarta Kumar, Aleem Siddiqui, and Nishant Goyal. "Neuropsychology of Prefrontal Cortex." *Indian Journal of Psychiatry.* Medknow Publications, July-Aug. 2008. http://www.ncbi.nlm.nih.gov/pmc/articles/PMC2738354/

26. Arnsten, Amy F. T. "Stress Signalling Pathways That Impair Prefrontal Cortex Structure and Function." *Nature Reviews. Neuroscience.* U.S. National Library of Medicine, June 2009. http://www.ncbi.nlm.nih.gov/pmc/articles/PMC2907136/#BX1

27. Taleb, Nassim Nicholas. (2014) *Antifragile: Things That Gain from Disorder (Incerto).* N.p.: Print.

28. Lieberman, D. The Story of the Human Body: Evolution, Health and Disease. (2013). Pantheon Books.

29. Gluckman, P. Hanson, M. Mismatch. (2006). Oxford University Press

30. Arnsten, Amy F. T. "Stress Signalling Pathways That Impair Prefrontal Cortex Structure and Function." *Nature Reviews. Neuroscience*. U.S. National Library of Medicine, June 2009. http://www.ncbi.nlm.nih.gov/pmc/articles/PMC2907136/#BX1

31. Dunn, Rob. "Human Ancestors Were Nearly All Vegetarians." *Scientific American Blog Network*. 23 July 2012. http://blogs.scientificamerican.com/guest-blog/human-ancestors-were-nearly-all-vegetarians/

32. Lum, Zi-Ann. "WATCH: It Takes 19 Ingredients To Make One McDonald's French Fry." *The Huffington Post*. 22 Jan. 2015. http://www.huffingtonpost.ca/2015/01/22/mcdonalds-french-fries-ingredients_n_6528848.html

33. Pontzer, Herman, David A. Raichlen, Brian M. Wood, Audax Z.P. Mabulla, Susan B. Racette, and Frank W. Marlowe. "Hunter-Gatherer Energetics and Human Obesity." *PLOS ONE:*. 25 July 2012. http://journals.plos.org/plosone/article?id=10.1371%2Fjournal.pone.0040503

34. Insel, Thomas. "Director's Blog: The Global Cost of Mental Illness." *NIMH RSS*. 28 Sept. 2011. http://www.nimh.nih.gov/about/director/2011/the-global-cost-of-mental-illness.shtml

35. Brun, John Pierre. "Work Related Stress: Scientific evidence-base of risk factors, prevention and costs." http://www.who.int/occupational_health/topics/brunpres0307.pdf

36. 36. *The Cost of Workplace Stress in Australia*. N.p.: *Medibank*. Aug. 2008. <http://www.medibank.com.au/client/documents/pdfs/the-cost-of-workplace-stress.pdf

37. Goh, Joel, Pfeffer J, and Zenios SA. "The Relationship Between Workplace Stressors and Mortality and Health Costs in the United States." Management Science: INFORMS. Management Science, n.d. Web. Mar. 2015. http://www.aepsal.com/wp-content/uploads/2015/03/EstresEnUSA.pdf

38. "PTSD Statistics." *PTSD United*.http://www.ptsdunited.org/ptsd-statistics-2/

39. "Facts & Statistics | Anxiety and Depression Association of America, ADAA." Anxiety and Depression Association of America, n.d. http://www.adaa.org/about-adaa/press-room/facts-statistics

40. Fang, Xiangming, Derek S. Brown, Curtis Florence, and James A. Mercy. "The Economic Burden of Child Maltreatment in the United States And Implications for Prevention." Child Abuse & Neglect. U.S. National Library of Medicine. Vol 36:2, Feb 2012:156–165 http://www.sciencedirect.com/science/article/pii/S0145213411003140

41. "Child Abuse and Neglect Cost the United States $124 Billion." Centers for Disease Control and Prevention, 1 Feb. 2012. http://www.cdc.gov/media/releases/2012/p0201_child_abuse.html

42. "Trends & Statistics." *Trends & Statistics*. 20 Aug. 2015. http://www.drugabuse.gov/related-topics/trends-statistics

43. "Emotion and Stress." *Neural Plasticity* 6.Supplement-1 (1999): 52-65. <http://www.parkerphd.com/PDFs/Chapt_09.pdf

44. Martin, Loren J. "Reducing Social Stress Elicits Emotional Contagion of Pain in Mouse and Human Strangers." 15 Jan. 2015. <http://www.cell.com/current-biology/abstract/S0960-9822(14)01489-4

45. Sapolsky, Robert M. "When Stress Rises, Empathy Suffers." *The Wall Street Journal.* 16 Jan. 2015. http://www.wsj.com/articles/when-stress-rises-empathy-suffers-1421423942

46. Damasio, A., and GB Carvalho. "The Nature of Feelings: Evolutionary and Neurobiological Origins." *National Center for Biotechnology Information.* U.S. National Library of Medicine, Feb. 2013. http://www.ncbi.nlm.nih.gov/pubmed/23329161

47. Damasio, A. "Feelings of Emotion and the Self." *National Center for Biotechnology Information.* U.S. National Library of Medicine, Oct. 2003. http://www.ncbi.nlm.nih.gov/pubmed/14625365

48. Damasio, AR. "The Somatic Marker Hypothesis and the Possible Functions of the Prefrontal Cortex." *National Center for Biotechnology Information.* U.S. National Library of Medicine, 29 Oct. 1996. http://www.ncbi.nlm.nih.gov/pubmed/8941953

49. LeDoux, Joseph E. *Feelings: What Are They & How Does the Brain Make Them?* n. pag. New York University. <http://www.cns.nyu.edu/home/LeDoux/pdf/daed_LeDoux_2015.pdf

50. LeDoux, Joseph. ""Inside Out" Draws from Darwin's Theory of the Nature of Emotions — I Think He Got It Wrong." *Salon.com.* 19 July 2015. http://www.salon.com/2015/07/19/inside_outs_psychology_of_emotion_is_based_on_the_darwinian_tradition_but_thats_just_one_view/

51. "Somatic Marker Hypothesis." *Wikipedia.* Wikimedia Foundation, 24 Dec. 2015. https://en.wikipedia.org/wiki/Somatic_marker_hypothesis

52. Lay, Keng-Ling, Everett Waters, German Posada, and Doreen Ridgeway. *Attachment Security, Affect Regulation, and Defensive Responses to Mood Induction.* Monographs of the Society for Research in Child Development. Vol. 60, No. 2/3, 1995. <http://

www.psychology.sunysb.edu/attachment/online/online_2/mood_induction.pdf

53. Gross, James J. "Emotion Regulation: Affective, Cognitive, and Social Consequences." n. pag. <https://sprweb.org/articles/Gross02.pdf

54. Al Turtle, personal communication.

55. Jones, David S., Scott H. Podolsky, and Jeremy A. Greene. "The Burden of Disease and the Changing Task of Medicine — NEJM." *New England Journal of Medicine*. 21 June 2012. http://www.nejm.org/doi/full/10.1056/NEJMp1113569

56. "Long-Term Diabetes." *Diabetes* 2.6 (1953): 500-01. *Centers for Disease Control and Prevention*. Oct. 2014/. <http://www.cdc.gov/diabetes/statistics/slides/long_term_trends.pdf

57. Salleh, Mohd. Razali. "Life Event, Stress and Illness." *The Malaysian Journal of Medical Sciences : MJMS*. Penerbit Universiti Sains Malaysia, Oct. 2008. http://www.ncbi.nlm.nih.gov/pmc/articles/PMC3341916/

58. "Healthy, United States, 2014." *Centers for Disease Control and Prevention*. 2014. <http://www.cdc.gov/nchs/data/hus/hus14.pdf#085

59. "OECD ILibrary: Statistics / Health at a Glance: Europe / 2012 /." *OECD ILibrary: Statistics / Health at a Glance: Europe 2012*. http://www.oecd-ilibrary.org/sites/9789264183896-en/03/11/index.html?itemId=%2Fcontent%2Fchapter%2F9789264183896-38-en&_csp_=45f4df11dc99cd2019aa9aa30865f74f

60. "Healthy, United States, 2014." *Centers for Disease Control and Prevention*. 2014. <http://www.cdc.gov/nchs/data/hus/hus14.pdf#085

61. Luthar, Suniya S., Samuel H. Barkin, and Elizabeth J. Crossman. ""I Can, Therefore I Must": Fragility in the Upper-middle Classes." *Dev Psychopathol Development and Psychopathology* 25.4pt2 (2013): 1529-549. <http://www.isacs.org/uploads/file/Annual%20Conference/Annual%202013/Luthar%20I%20think.pdf

62. Elzinga, B.M., and K. Roelofs. "Cortisol-induced Impairments of Working Memory Require Acute Sympathetic Activation." Mar. 2005. <https://www.researchgate.net/profile/K_Roelofs/publication/8005147_Cortisol-induced_impairments_of_working_memory_require_acute_sympathetic_activation/links/0912f50c07dc32649e000000.pdf

63. Elzinga, B. M., and K. Roelofs. "Cortisol-induced Impairments of Working Memory Require Acute Sympathetic Activation." National Center for Biotechnology Information. Behav Neurosci. 2005 Feb;119(1):98-103

64. Arnsten, Amy F. T. "Stress Signalling Pathways That Impair Prefrontal Cortex Structure and Function." *Nature Reviews. Neuroscience.* U.S. National Library of Medicine, June 2009. http://www.ncbi.nlm.nih.gov/pmc/articles/PMC2907136/

65. Otto, A. R., C. M. Raio, A. Chiang, E. A. Phelps, and N. D. Daw. "Working-memory Capacity Protects Model-based Learning from Stress." *Proceedings of the National Academy of Sciences* 110.52 (2013): 20941-0946. 11 Nov. 2013. <http://www.pnas.org/content/110/52/20941.full.pdf

66. Wilkinson, Richard G. "Socioeconomic Determinants of Health: Health Inequalities: Relative or Absolute Material Standards?" *The BMJ.* 22 Feb.1997. http://www.bmj.com/content/314/7080/591?ijkey=7a69310536aa0317f521ad6df8b4669ad1199e56&keytype2=tf_ipsecsha

67. Timmer, Marjan. "Neurobiological Mechanisms Involved in the Establishment and Maintenance of Dominance Hierarchies and Its Modulation by Stress in Rats." 2011. <http://infoscience.epfl.ch/record/152305/files/EPFL_TH4892.pdf?version=1

68. Cordero, Maria Isabel, and Carmen Sandi. "Stress Amplifies Memory for Social Hierarchy." <http://infoscience.epfl.ch/record/125751/files/LGC_Cordero_I.pdf?version=2

69. Gesquiere, Laurence R., Niki H. Learn, M. Carolina M. Simao, Patrick O. Onyango, Susan C. Alberts, and Jeanne Altmann. "Life at the Top: Rank and Stress in Wild Male Baboons." *Science (New York, N.Y.).* U.S. National Library of Medicine, 15 July 2011. http://www.ncbi.nlm.nih.gov/pmc/articles/PMC3433837/

70. Ozbay, Fatih, Douglas C. Johnson, Eleni Dimoulas, C.A. Morgan, Dennis Charney, and Steven Southwick. "Social Support and Resilience to Stress: From Neurobiology to Clinical Practice." *Psychiatry (Edgmont).* Matrix Medical Communications, May 2007. http://www.ncbi.nlm.nih.gov/pmc/articles/PMC2921311/

71. Gächter, Martin, David A. Savage, and Benno Torgler. "Policing: An International Journal of Police Strategies & Management." *The Relationship between Stress, Strain and Social Capital: Vol 34, No 3.* 1997. http://www.emeraldinsight.com/doi/abs/10.1108/13639511111157546?journalCode=pijpsm

72. "Social Capital." *Wikipedia.* Wikimedia Foundation, 2 Mar. 2016. https://en.wikipedia.org/wiki/Social_capital

73. Putnam, Robert D. *Bowling Alone.* Köln: Janus-Verl.-Ges., 2002. Jan. 1995. <http://archive.realtor.org/sites/default/files/BowlingAlone.pdf

74. Weil, Frederick, Matthew R. Lee, and Edward S. Shihadeh. "The Burdens of Social Capital: How Socially-involved People Dealt with Stress after Hurricane Katrina." *Science Direct.*

June 2011. http://www.sciencedirect.com/science/article/pii/S0049089X11001104

75. Kawachi, I., B. P. Kennedy, K. Lochner, and D. Prothrow-Stith. "Social Capital, Income Inequality, and Mortality." *American Journal of Public Health.* U.S. National Library of Medicine, Sept. 1997. http://www.ncbi.nlm.nih.gov/pmc/articles/PMC1380975/?page=1

76. Pearce, Neil, and George Davey Smith. "Is Social Capital the Key to Inequalities in Health?" *American Journal of Public Health.* © American Journal of Public Health 2003, Jan. 2003. http://www.ncbi.nlm.nih.gov/pmc/articles/PMC1447706/

77. "List of Presidents of Ecuador." *Ecuaworld.*http://www.ecuaworld.com/ecuador-information-center/list-presidents-ecuador/

78. "Constitutional History of Ecuador." *Wikipedia.* Wikimedia Foundation, 25 Dec. 2015. https://en.wikipedia.org/wiki/Constitutional_history_of_Ecuador

79. "Currency of Ecuador." *Wikipedia.* Wikimedia Foundation, 23 Feb. 2016. https://en.wikipedia.org/wiki/Currency_of_Ecuador

80. D'Argenio, Valeria, and Francesco Salvatore. "The Role of the Gut Microbiome in the Healthy Adult Status." Jan. 2015. http://www.sciencedirect.com/science/article/pii/S0009898115000170

81. Zatorski, Hubert, and Jakub Fichna. "What Is the Future of the Gut Microbiota-Related Treatment? Toward Modulation of Microbiota in Preventive and Therapeutic Medicine." Frontiers in Medicine. Frontiers Media S.A. 10 Jul 2014.

82. D'Argenio, Valeria, and Francesco Salvatore. "The Role of the Gut Microbiome in the Healthy Adult Status." Jan. 2015. http://www.sciencedirect.com/science/article/pii/S0009898115000170

83. Moschen, Alexander R., Verena Wieser, and Herbert Tilg. "Dietary Factors: Major Regulators of the Gut's Microbiota." 7 Aug. 2012. http://www.ncbi.nlm.nih.gov/pmc/articles/PMC3493718/

84. Mayer, E.A, K. Tillisch, and A. Gupta. "Gut/brain Axis and the Microbiota." *National Center for Biotechnology Information. U.S. National Li*brary of Medicine, 2 Mar. 2015. http://www.ncbi.nlm.nih.gov/pubmed/25689247

85. Oyri, S.F., G. Muzes, and F. Sipos. "Dysbiotic Gut Microbiome: A Key Element of Crohn's Disease." *National Center for Biotechnology Information.* U.S. National Library of Medicine, Dec. 2015. http://www.ncbi.nlm.nih.gov/pubmed/26616659

86. Schnabl, Bernd, and David A. Brenner. "Interactions Between the Intestinal Microbiome and Liver Diseases." *Science Direct.* Jan. 2014. <http://www.sciencedirect.com/science/article/pii/S0016508514000778

87. Severance, Emily G., Robert H. Yolken, and William W. Eaton. "Autoimmune Diseases, Gastrointestinal Disorders and the Microbiome in Schizophrenia: More than a Gut Feeling." *Schizophrenia Research.* <http://www.schres-journal.com/article/S0920-9964(14)00319-3/abstract

88. Le Chatelier, Emmanuelle. "Richness of Human Gut Microbiome Correlates with Metabolic Markers." *Nature.com.* 10 Apr. 2012. <http://www.nature.com/nature/journal/v500/n7464/full/nature12506.html

89. Wang, Yunwei, and Jeanette D. Hoenig. "16S RRNA Gene-based Analysis of Fecal Microbiota from Preterm Infants with and without Necrotizing Enterocolitis." *Nature.com.* Nature Publishing Group, 16 Apr. 2009. http://www.nature.com/ismej/journal/v3/n8/full/ismej200937a.html

90. Bisgaard, H., N. Li, K. Bonnelykke, BL Chawes, T. Skov, G. Paludan-Müller, J. Stokholm, B. Smith, and KA Krogfelt. "Reduced Diversity of the Intestinal Microbiota during Infancy Is Associated with Increased Risk of Allergic Disease at School Age." *National Center for Biotechnology Information*. U.S. National Library of Medicine, Sept. 2011. http://www.ncbi.nlm.nih.gov/pubmed/21782228

91. Lopes, C., J. Soares, F. Tavaria, A. Duarte, O. Correia, and O. Sokhatska. "Chitosan Coated Textiles May Improve Atopic Dermatitis Severity by Modulating Skin Staphylococcal Profile: A Randomized Controlled Trial." *National Center for Biotechnology Information*. U.S. National Library of Medicine, 30 Nov. 2015. http://www.ncbi.nlm.nih.gov/pubmed/26618557

92. Szekeres, M., F. Somogyvári, K. Bencsik, Z. Szolnoki, L. Vécsei, and Y. Mandi. "GENETIC POLYMORPHISMS OF HUMAN β-DEFENSINS IN PATIENTS WITH MULTIPLE SCLEROSIS." *National Center for Biotechnology Information*. U.S. National Library of Medicine, 30 Mar. 2015. http://www.ncbi.nlm.nih.gov/pubmed/26434201

93. Riccio, P., and R. Rossano. "Nutrition Facts in Multiple Sclerosis." *National Center for Biotechnology Information*. U.S. National Library of Medicine, 18 Feb. 2015. http://www.ncbi.nlm.nih.gov/pubmed/25694551

94. Tomasello, G., P. Tralongo, F. Amoroso, P. Damiani, E. Sinagra, and M. Noto. "DYSMICROBISM, INFLAMMATORY BOWEL DISEASE AND THYROIDITIS: ANALYSIS OF THE LITERATURE." *National Center for Biotechnology Information*. U.S. National Library of Medicine, Apr.-May 2015. http://www.ncbi.nlm.nih.gov/pubmed/26122213

95. Torres, N., M. Guevara-Cruz, LA Velázquez-Villegas, and AR Tovar. "Nutrition and Atherosclerosis." *National Center for*

Biotechnology Information. U.S. National Library of Medicine, July 2015. http://www.ncbi.nlm.nih.gov/pubmed/26031780

96. Drosos, I., Tavridou A, and Kolios G. "New Aspects on the Metabolic Role of Intestinal Microbiota in the Development of Atherosclerosis." *National Center for Biotechnology Information*. U.S. National Library of Medicine, Apr. 2015. http://www.ncbi.nlm.nih.gov/pubmed/25676802

97. Anahtar, MN. "Cervicovaginal Bacteria Are a Major Modulator of Host Inflammatory Responses in the Female Genital Tract." *National Center for Biotechnology Information*. U.S. National Library of Medicine, 19 May 2015. http://www.ncbi.nlm.nih.gov/pubmed/25992865

98. Sze, MA, PA Dimitriu, and M. Suzuki. "Host Response to the Lung Microbiome in Chronic Obstructive Pulmonary Disease." *National Center for Biotechnology Information*. U.S. National Library of Medicine, 15 Aug. 2015. http://www.ncbi.nlm.nih.gov/pubmed/25945594

99. Yang, T., V. Rodriguez, MM Santisteban, and E. Li. "Gut Dysbiosis Is Linked to Hypertension." *National Center for Biotechnology Information*. U.S. National Library of Medicine, June 2015. http://www.ncbi.nlm.nih.gov/pubmed/25870193

100. Zapata, HJ, and VJ Quagliarello. "The Microbiota and Microbiome in Aging: Potential Implications in Health and Age-related Diseases." *National Center for Biotechnology Information*. U.S. National Library of Medicine, Apr. 2015. http://www.ncbi.nlm.nih.gov/pubmed/25851728

101. Hoen, AG, J. Li, LA Moulton, GA O'Toole, ML Housman, and DC Koestler. "Associations between Gut Microbial Colonization in Early Life and Respiratory Outcomes in Cystic Fibrosis." *National Center for Biotechnology Information*. U.S. National

Library of Medicine, July 2015. http://www.ncbi.nlm.nih.gov/pubmed/25818499

102. Zhang, H., X. Liao, JB Sparks, and XM Luo. "Dynamics of Gut Microbiota in Autoimmune Lupus." *National Center for Biotechnology Information*. U.S. National Library of Medicine, Dec. 2014. http://www.ncbi.nlm.nih.gov/pubmed/25261516

103, 104. Cho, Ilseung, and Martin J. Blaser. "The Human Microbiome: At the Interface of Health and Disease." *Nature Reviews. Genetics*. U.S. National Library of Medicine, 13 Mar. 2012. http://www.ncbi.nlm.nih.gov/pmc/articles/PMC3418802/

105. Turnbaugh, Peter J., Vanessa K. Ridaura, Jeremiah J. Faith, Frederico E. Rey, Rob Knight, and Jeffrey I. Gordon. "The Effect of Diet on the Human Gut Microbiome." 11 Nov. 2009. <http://stm.sciencemag.org/content/1/6/6ra14.short

106. Shanle, Erin K., and Wei Xu. "Endocrine Disrupting Chemicals Targeting Estrogen Receptor Signaling: Identification and Mechanisms of Action." Chemical Research in Toxicology. U.S. National Library of Medicine. 11 Jan 2014.

107. Strachan, David P. "Family Size, Infection and Atopy: The First Decade of the 'hygiene Hypothesis'" http://thorax.bmj.com/content/55/suppl_1/S2

108. Raison CL, Lowry CA, Rook GAW. Inflammation, sanitation and consternation: loss of contact with co-evolved, tolerogenic micro-organisms and the pathophysiology and treatment of major depression" Arch Gen Psychiatry2010;67(12) 1211-24.

109. Rook GAW, Lowry CA, Raison CL. Microbial Old Friends, immunoregulation and stress resilience. Evolution, Medicine and Public Health. 2013: 46-64.

110. Rook, G. A. W.; Lowry, C. A.; Raison, C. L. (2013). "Microbial 'Old Friends', immunoregulation and stress resilience". Evolution, Medicine, and Public Health 2013: 46–64. doi:10.1093/emph/eot004.

111. Rook GAW, Dalgleish A (2011). "Infection, immunoregulation and cancer". Immunological Reviews 240: 141–59.

112. Janssen, Sarah, Gina Solomon, and Ted Schettler. "Chemical Contaminants and Human Disease: A Summary of Evidence." <http://c.ymcdn.com/sites/www.njsna.org/resource/resmgr/imported/Chemical%20Contaminants%20and%20Human%20Disease%20-%20A%20Summary%20of%20Evide.pdf

113. "Highly Processed Foods Linked to Addictive Eating | University of Michigan News." *Michigan News*. University of Michigan, 18 Feb. 2015. http://ns.umich.edu/new/releases/22693-highly-processed-foods-linked-to-addictive-eating

114. Lustig, Robert H., Kathleen Mulligan, Susan M. Noworolski, Viva W. Tai, Michael J. Wen, Ayca Erkin-Cakmak, Alejandro Gugliucci, and Jean-Marc Schwarz. "Isocaloric Fructose Restriction and Metabolic Improvement in Children with Obesity and Metabolic Syndrome." *Wiley Online Library*. 26 Oct. 2015. http://onlinelibrary.wiley.com/doi/10.1002/oby.21371/abstract

115. Park, Alice. "Sugar Is Toxic, Says New Study." Time, 28 Oct. 2015. http://time.com/4087775/sugar-is-definitely-toxic-a-new-study-says/

116. "Pollution." *Wikipedia*. Wikimedia Foundation, 15 Feb. 2016. https://en.wikipedia.org/wiki/Pollution

117. Lang, Susan S. "Water, Air and Soil Pollution Causes 40 Percent of Deaths Worldwide, Cornell Research Survey Finds

| Cornell Chronicle." *Cornell Chronicle*. Cornell University, 2 Aug. 2007. http://www.news.cornell.edu/stories/2007/08/pollution-causes-40-percent-deaths-worldwide-study-finds

118. Kampa, Marilena, and Elias Castanas. "Human Health Effects of Air Pollution." *Science Direct*. June 2007. http://www.sciencedirect.com/science/article/pii/S0269749107002849

119. "Burden of Disease from Ambient and Household Air Pollution." *World Health Organization*.http://www.who.int/phe/health_topics/outdoorair/databases/en/

Made in the USA
Monee, IL
23 July 2020